FOUNDATIONS:
LOGIC, LANGUAGE, AND MATHEMATICS

FOUNDATIONS: LOGIC, LANGUAGE, AND MATHEMATICS

Edited by

HUGUES LEBLANC, ELLIOTT MENDELSON,

and

ALEX ORENSTEIN

Reprinted from
Synthese, Vol. 60 Nos. 1 and 2

D. REIDEL PUBLISHING COMPANY

DORDRECHT / BOSTON

TABLE OF CONTENTS

CITY COLLEGE STUDIES
IN THE HISTORY AND PHILOSOPHY
OF SCIENCE AND TECHNOLOGY:
SERIES EDITORS' PREFACE

Recent years have seen the emergence of several new approaches to the history and philosophy of science and technology. For one, what were perceived by many as separate, though perhaps related, fields of inquiry have come to be regarded by more and more scholars as a single discipline with different areas of emphasis. In this discipline any profound understanding so deeply intertwines history and philosophy that it might be said, to paraphrase Kant, that philosophy without history is empty and history without philosophy is blind.

Another contemporary trend in the history and philosophy of science and technology has been to bring together the English-speaking and continental traditions in philosophy. The views of those who do analytic philosophy and the views of the hermeneuticists have combined to influence the thinking of some philosophers in the English-speaking world, and over the last decade that influence has been felt in the history and philosophy of science. There has also been the long-standing influence in the West of continental thinkers working on problems in the philosophy of technology. This synthesis of two traditions has made for a richer fund of ideas and approaches that may change our conception of science and technology.

Still another trend that is in some ways a combination of the previous two consists of the work of those characterized by some as the "Friends of Discovery" and by others as the bringers of the "New Fuzziness". This approach to the history and philosophy of science and technology concentrates on *change*, *progress*, and *discovery*. It has raised old epistemological questions under the guise of the problem of rationality in the sciences. Although this approach has its origins in the work of Thomas Kuhn in the United States, attempts to express his ideas in explicit set-theoretical or model-theoretic terms are now centered in Germany.

Synthese **60** (1984) 1.

The more traditional approaches to the history and philosophy of science and technology continue as well, and probably will continue as long as there are skillful practitioners such as Carl Hempel, Ernest Nagel, and their students.

Finally, there are still other approaches that address some of the technical problems arising when we try to provide an account of belief and of rational choice. – These include efforts to provide logical frameworks within which we can make sense of these notions.

This series will attempt to bring together work from all of these approaches to the history and philosophy of science and technology in the belief that each has something to add to our understanding.

The volumes of this series have emerged either from lectures given by authors while they served as honorary visiting professors at the City College of New York or from conferences sponsored by that institution. The City College Program in the History and Philosophy of Science and Technology oversees and directs these lectures and conferences with the financial aid of the Association for Philosophy of Science, Psychotheraphy, and Ethics.

MARTIN TAMNY
RAPHAEL STERN

PREFACE

The papers in this collection stem largely from the conference 'Foundations: Logic, Language, and Mathematics' held at the Graduate Center of the City University of New York on 14–15 November 1980. The conference was sponsored by the Philosophy Program at the Graduate Center of the City University of New York and the Association for Philosophy of Science, Psychotherapy, and Ethics. We wish to express our gratitude and appreciation to these organizations and to thank the series editors, Raphael Stern and Martin Tamny, for their help with the conference and the preparation of this collection.

HUGUES LEBLANC
ELLIOTT MENDELSON
ALEX ORENSTEIN

MELVIN FITTING

A SYMMETRIC APPROACH TO AXIOMATIZING QUANTIFIERS AND MODALITIES

1. INTRODUCTION

We present an axiomatization of several of the basic modal logics, with the idea of giving the two modal operators $\Box$ and $\Diamond$ equal weight as far as possible. Then we present a parallel axiomatization of classical quantification theory, working our way up through a sequence of rather curious subsystems. It will be clear at the end that the essential difference between quantifiers and modalities is amusing in a vacuous sort of way. Finally we sketch tableau proof systems for the various logics we have introduced along the way. Also, the "natural" model theory for the subsystems of quantification theory that come up is somewhat curious. In a sense, it amounts to a "stretching out" of the Henkin-style completeness proof, severing the maximal consistent part of the construction quite thoroughly from the part of the construction that takes care of existential-quantifier instances.

2. BACKGROUND

It is contrary to the spirit of what we are doing to take one modal operator as primitive and define the other, or, one quantifier as primitive and define the other. So, by the same token, we take as primitive all the standard propositional connectives too. Thus, we have available all of $\wedge$, $\vee$, $\sim$, $\supset$, $\Box$, $\Diamond$, $\forall$, $\exists$. We also take as primitive a truth constant $\top$ and a falsehood constant $\bot$.

For our treatment of propositional modal logic we assume we have a countable list of atomic formulas, and that the set of formulas is built up from them in the usual way. We will use the letters "X", "Y", etc., to denote such formulas.

For quantification theory, we assume we have a countable list of variables and also a disjoint countable list of parameters. We will use the letters "x", "y", etc., to denote variables, and "a", "b", etc., to denote parameters. Formulas are built up in the usual way, with the understanding that a sentence contains no free variables, though it may

Synthese **60** (1984) 5–19.   0039–7857/84/0601–0005 $01.50

contain parameters. We follow the convention that if $\varphi(x)$ is a formula with only x free, then $\varphi(a)$ is the result of replacing all free occurrences of x by occurences of a in φ.

We will use Kripke's model theory for modal logics (and an analog for quantificational theories, to be described in Section 4). For us, a *Kripke model* is a quadruple $\langle \mathcal{G}, \Phi, \mathcal{R}, \Vdash \rangle$ where: (1) $\mathcal{G}$ is a nonempty set (of possible worlds); (2) Φ is a possible empty subset of $\mathcal{G}$ (of so-called queer worlds); (3) $\mathcal{R}$ is a relation on $\mathcal{G}$ (of accessibility); and (4) $\Vdash$ is a relation between members of $\mathcal{G}$ and formulas meeting the following conditions (we write $\Gamma \nVdash X$ as short for not $- \Gamma \Vdash X$):

for every $\Gamma \in \mathcal{G}$

(a)	$\Gamma \Vdash \top$ and $\Gamma \nVdash \bot$;
(b)	$\Gamma \Vdash (X \wedge Y)$ iff $\Gamma \Vdash X$ and $\Gamma \Vdash Y$;
(c)	$\Gamma \Vdash (X \vee Y)$ iff $\Gamma \Vdash X$ or $\Gamma \Vdash Y$;
(d)	$\Gamma \Vdash \sim X$ iff $\Gamma \nVdash X$;
(e)	$\Gamma \Vdash (X \supset Y)$ iff $\Gamma \nVdash X$ or $\Gamma \Vdash Y$;

for every $\Gamma \in \Phi$

(f)	$\Gamma \Vdash \Diamond X$ but $\Gamma \nVdash \Box X$;

for every $\Gamma \in \mathcal{G} - \Phi$

(g)	$\Gamma \Vdash \Box X$ iff for every $\Delta \in \mathcal{G}$ such that $\Gamma \mathcal{R} \Delta$, $\Delta \Vdash X$;
(h)	$\Gamma \Vdash \Diamond X$ iff for some $\Delta \in \mathcal{G}$ such that $\Gamma \mathcal{R} \Delta$, $\Delta \Vdash X$.

Thus, in a queer world, anything is possible, nothing is necessary. But otherwise, necessary truth means truth in all accessible worlds. A Kripke model in which $\Phi = \phi$ is called *normal*. See Kripke [2] and [3].

We also will be making much use of *uniform notation*, as introduced in Smullyan [6]. For this purpose, the nonatomic formulas of propositional logic are grouped into α-formulas (those that act conjuctively) and β-formulas (those that act disjunctively). These, and their *components*, α_1, α_2 and β_1, β_2, respectively, are presented in the following charts:

α	α_1	α_2		β	β_1	β_2
$(X \wedge Y)$	X	Y		$(X \vee Y)$	X	Y
$\sim(X \vee Y)$	$\sim X$	$\sim Y$		$\sim(X \wedge Y)$	$\sim X$	$\sim Y$
$\sim(X \supset Y)$	X	$\sim Y$		$(X \supset Y)$	$\sim X$	Y
$\sim \sim X$	X	X				

Notice that (b)–(e) in the definition of a Kripke model above gives us

$$\Gamma \Vdash \alpha \text{ iff } \Gamma \Vdash \alpha_1 \text{ and } \Gamma \Vdash \alpha_2$$
$$\Gamma \Vdash \beta \text{ iff } \Gamma \Vdash \beta_1 \text{ or } \Gamma \Vdash \beta_2.$$

Indeed, this could be used instead of (b)–(e).

We extend this uniform notation to the modal operators (as in Fitting [1]) by defining the ν-formulas (necessaries) and π-formulas (possibles) and their components ν_0 and π_0, respectively, as follows:

ν	ν_0		π	π_0
$\Box X$	X		$\Diamond X$	X
$\sim \Diamond X$	$\sim X$		$\sim \Box X$	$\sim X$

Notice that (f)–(h) in the definition of a Kripke model above, taken together with (d), give us the following equivalent conditions:

for every $\Gamma \in \Phi$,

(f′) $\Gamma \Vdash \pi$ but $\Gamma \nVdash \nu$;

for every $\Gamma \in \mathcal{G} - \Phi$

(g′) $\Gamma \Vdash \nu$ iff for every $\Delta \in \mathcal{G}$ such that $\Gamma \mathcal{R} \Delta$, $\Delta \Vdash \nu_0$;

(h′) $\Gamma \Vdash \pi$ iff for some $\Delta \in \mathcal{G}$ such that $\Gamma \mathcal{R} \Delta$, $\Delta \Vdash \pi_0$.

Finally we present uniform notation for the quantifiers, as in Smullyan [6] or as more fully developed, Smullyan [7]. The quantified sentences are divided into γ-sentences (universals) and δ-sentences (existentials). This time, rather than components, we have *instances*, $\gamma(a)$ and $\delta(a)$, one for each parameter a.

γ	$\gamma(a)$		δ	$\delta(a)$
$(\forall x)\,\phi(x)$	$\phi(a)$		$(\exists x)\,\phi(x)$	$\phi(a)$
$\sim(\exists x)\,\phi(x)$	$\sim \phi(a)$		$\sim(\forall x)\phi(x)$	$\sim \phi(a)$

The idea is, in a classical first-order model in which the domain consists of the set of parameters, and each parameter is interpreted as naming itself:

γ is true iff for all a, $\gamma(a)$ is true;

δ is true iff for some a, $\delta(a)$ is true.

So much for background.

3. MODAL LOGICS, AXIOMATICALLY

Let us assume that we have an axiomatization of the classical pro-positional calculus, with modus ponens as the only rule. We will build on that in our introduction of rules and axioms pertinent to the modal operators.

Modal logics are often formulated with a rule of necessitation, which we could give as

$$\frac{\nu_0}{\nu}.$$

But this gives a central role to the ν-formulas and, for no reason other than autocratic whim, we want to develop modal logic as far as possible giving equal weight to both ν-and π-formulas, giving to both neces-saries and possibles a fair share. After a certain amount of experimen-tation we hit on the following curious rule:

$$(M) \qquad \frac{\pi_0 \vee \nu_0}{\pi \vee \nu}$$

(we use M for modalization). As a matter of fact, rule M is a correct rule of inference, as the following argument shows.

Suppose $\langle \mathcal{G}, \Phi, \mathcal{R}, \Vdash \rangle$ is a Kripke model, and $\pi_0 \vee \nu_0$ is valid in it (holds at every possible world). Let $\Gamma \in \mathcal{G}$; we show $\Gamma \Vdash \pi \vee \nu$. Well, if $\Gamma \in \Phi$, that is, if Γ is queer, then $\Gamma \Vdash \pi$; so, trivially, $\Gamma \Vdash \pi \vee \nu$. Otherwise, $\Gamma \in \mathcal{G} - \Phi$. Now suppose $\Gamma \nVdash \pi$. Then it must be that in every Δ such that $\Gamma \mathcal{R} \Delta$, we have $\Delta \nVdash \pi_0$. But $\pi_0 \vee \nu_0$ is assumed valid in our model, so it holds at every such Δ. Thus for every Δ such that $\Gamma \mathcal{R} \Delta$, $\Delta \Vdash \nu_0$; hence $\Gamma \Vdash \nu$. Thus again $\Gamma \Vdash \pi \vee \nu$.

Rule M (which is actually four rules when translated out of uniform notation) is quite a useful rule. It is rather easy to show that, by using it, one may derive the usual interdefinitions of the modal operators (as mutual equivalences). Also one may show the following are derived rules:

$$\frac{X \supset Y}{\Box X \supset \Box Y} \qquad \frac{X \supset Y}{\Diamond X \supset \Diamond Y}.$$

Then one may show, in the usual way, that replacement of proved equivalences holds as a derived rule:

$$\frac{X \equiv X'}{Z \equiv Z'}$$

where Z' results from Z by replacing some occurrences of X in Z by occurrences of X'. Here we have used $\equiv$ as an abbreviation for mutual implication. (Actually a stronger version may be shown, concerning "semisubstitutivity" of implication, in which we must take into account the positiveness and negativeness of occurrences as well. The details needn't concern us here.)

Now go back and look again at the justification we gave for rule M. In the argument that $\pi \vee \nu$ held at the nonqueer world Γ we never needed that $\pi_0 \vee \nu_0$ held at *every* world of the model; we only needed that it held at every world accessible from Γ. But to say that $\pi_0 \vee \nu_0$ holds at every world accessible from Γ is to say that $\Box(\pi_0 \vee \nu_0)$ holds at Γ. Thus, the same argument also shows the validity in all Kripke models of the schema

(M1) $\Box[\pi_0 \vee \nu_0] \supset [\pi \vee \nu]$.

Let us add it as an axiom schema then.

If we do so, it is not hard to show that we have a complete axiomatic counterpart of the Kripke model theory as given in section 2. That is, X is provable in the axiomatic system just described iff X holds at every world of every Kripke model. To show this, one may show completeness directly, using the now-common Lindenbaum-style construction, or one may show this axiom system is of equal strength with one standard in the literature, and rely on known completeness results. We skip the arguments.

The logic axiomatically characterized thus far is called C in Segerberg [5]. It is the smallest *regular* modal logic (see Segerberg [5]).

Now, in the model theory for C, one can have queer worlds in which everything is possible but nothing is necessary. Also there are no special conditions placed on the accessibility relation $\mathscr{R}$, so there can be "dead-end" worlds, worlds from which no world is accessible. In such a world, everything is necessary, nothing is possible. So, our next item of business is to rule out such strange worlds.

Worlds in which nothing is necessary may be eliminated by postulating that something is necessary. Let us, then, add the axiom

(M2) $\Box\top$.

The model theory appropriate to this (with respect to which one can prove completeness) is all Kripke models in which there are no queer worlds, that is, all normal Kripke models. The logic axiomatized thus far is the smallest *normal* logic, and is usually called K (see Segerberg [5]).

Next, worlds in which nothing is possible may be eliminated by postulating that something is possible. We take as an axiom

(M3) $\Diamond\top$.

The model theory appropriate to this is all normal Kripke models in which every world has some world accessible to it. The logic is generally called D. (Again, see Segerberg [5]).

Finally, we might want to restrict our attention to models in which each (normal) world is accessible to itself (in which the accessibility relation is reflexive). To do this one adds either (or both) of

(M4) $\nu \supset \nu_0$
$\quad \pi_0 \supset \pi$

The logic thus characterized is T.

Note that $\top \supset \Diamond\top$ is an instance of the second of these schemas, and since $\top$ is a tautology, $\Diamond\top$ follows by modus ponens. Thus with M4 added, M3 becomes redundant.

REMARKS. One goes from C to K by adding $\Box\top$ as an axiom. It is of some interest to consider a kind of halfway point, where instead of adding $\Box\top$ as an *assumed truth*, we take it as a *hypothesis*. That is, form the set S of formulas X such that $\Box\top \supset X$ is provable in C. This set S is, itself, a (rather strangely defined) modal logic, intermediate between C and K. It is closed under modus ponens, but not under rule M. Rather, it is closed under the weaker rule.

$$\frac{\Box(\pi_0 \vee \nu_0)}{\Box(\pi \vee \nu)},$$

which is equivalent to Becker's rule. It is, in fact, the logic axiomatized in Lemmon [4] as P2, but without his axiom $\Box X \supset X$ (our M4).

If we strengthen things a bit, by considering those X such that $\Box T \supset X$ is provable in C plus axiom M4, one gets the Lewis system S2. (Again, see Segerberg [5], chapter four.)

One can also play similar games with axiom M3 ($\Diamond T$) to produce interesting logics. We know very little about them.

4. QUANTIFIED LOGICS, AXIOMATICALLY

There is an obvious analogy (of sorts) between the modal operators and quantifiers. What we do in this section is to parallel the development of section 3, substituting quantifiers for the modal operators to see how far the analogy extends when things are done the way we did. The idea is simple: $(\forall x)$ may behave like $\Box$, $(\exists x)$ like $\Diamond$, γ like ν and δ like π. This may be so; we will see.

The language now is first order. Once again we assume a propositional-logic base with modus ponens as the sole rule.

First, the analog of rule M is

rule (Q) $\dfrac{\delta(a) \vee \gamma(a)}{\delta \vee \gamma}$

And, as a matter of fact, this is a correct rule of inference in classical first-order logic. This argument is left to the reader.

Using rule Q, one may show analogs of the results listed in section 3 based on rule M. Thus, one may derive the usual inter definability of the quantifiers (again, as mutual implication), and one may show that replacement of proved equivalences holds as a derived rule.

Now, if you actually thought through the "justification" of rule Q, almost certainly you also showed the validity, in all first-order models, of the schema

(Q1) $(\forall x)[\delta(x) \vee \gamma(x)] \supset [\delta \vee \gamma]$.

So let us add it as an axiom schema. We might, by analogy, call the resulting logic QC. It thus has rule Q and schema Q1.

The following is a reasonable question: What is an adequate model theory for QC, one with respect to which completeness can be shown? Well, the following rather curious one will do. We simply translate the corresponding modal model theory, making suitable adjustments to take care of the fact that quantified sentences have many instances, but modalized formulas have single components. We have chosen, for simplicity, to leave out any mention of the notion of an *interpretation* in a model. A more elaborate treatment would have to include it, but the following is enough for our purposes.

A *model* is a quintuple $\langle \mathcal{G}, \Phi, \mathcal{P}, \mathcal{R}, \Vdash \rangle$ where: (1) $\mathcal{G}$ is a nonempty set (of possible worlds); (2) $\Phi \subseteq \mathcal{G}$; (3) $\mathcal{P}$ is a mapping from members of $\mathcal{G}$ to nonempty sets of parameters; (4) $\mathcal{R}$ is a relation on $\mathcal{G}$; and (5) $\Vdash$ is a relation between members of $\mathcal{G}$ and sentences such that

for every $\Gamma \in \mathcal{G}$,

> conditions (a)–(e) as in section 1;

for every $\Gamma \in \Phi$,

> (f) $\Gamma \Vdash (\exists x)\varphi(x)$ but $\Gamma \nVdash (\forall x)\varphi(x)$;

for every $\Gamma \in \mathcal{G} - \Phi$

> (g) $\Gamma \Vdash (\forall x)\varphi(x)$ iff for every $\Delta \in \mathcal{G}$ such that $\Gamma \mathcal{R} \Delta$, and for every $a \in \mathcal{P}(\Delta)$, $\Delta \Vdash \varphi(a)$

> (h) $\Gamma \Vdash (\exists x)\varphi(x)$ iff for some $\Delta \in \mathcal{G}$ such that $\Gamma \mathcal{R} \Delta$, and for some $a \in \mathcal{P}(\Delta)$, $\Delta \Vdash \varphi(a)$.

Now, a sentence X is a theorem of the logic QC iff X holds at every world of every such model. We leave the correctness half to the reader. Note that if $\varphi(a)$ is provable in QC, so is the parameter variant $\varphi(b)$. This will be of use in proving correctness. And we briefly sketch the completeness half in the next section.

We note that we could restrict models so that, for each world Γ, $\mathcal{P}(\Gamma)$ is a singleton. It makes no difference. Now we can continue with our development, paralleling that of section 3.

In the quantifier models above there can be "queer" worlds (members of φ) in which everything exists, and hence no universal sentences hold. Such anomalies can be eliminated by adding the postulate

> (Q2) $(\forall x)\top$.

Doing so gives us a logic we may call QK. An adequate model theory for it is one that consists of all models of the sort described above, but with Φ always empty, that is no "queer" worlds.

Next, there may still be worlds from which no world is accessible. In such a world, every universal sentence holds, but no existential. They behave rather like empty-domain models of first-order logic. They may be ruled out by adding the axiom

(Q3) $(\exists x)\top$.

Let us call the resulting logic QD. Its model theory is that of QK with the additional requirement that for every world Γ there must be some world Δ such that $\Gamma \mathscr{R} \Delta$.

Finally (?) we may add an analog to M4, namely

(Q4) $\gamma \supset \gamma(a)$ or
          $\delta(a) \supset \delta$ (or both).

As might be expected, Q3 then becomes redundant. We can call the logic thus axiomatized QT. Its model theory is that of QD with the restriction that the accessibility relation $\mathscr{R}$ be reflexive.

We have reached the end of our parallel development (obviously, since we have matched everything we did in section 3). But we do *not* yet have the usual first-order logic. One doesn't want classical first-order models with lots of possible worlds in them: a classical model should be a one-world model.

Now, to only consider one-world Kripke modal models is to trivialize modal logic; it renders the modal operators useless. Necessary truth becomes the same thing as truth. That is, $A \supset \Box A$ is valid in all one-world models. Of course, this is not desirable. Modal operators are supposed to do something; they are supposed to have an effect; they ought not be vacuous.

Well, it is precisely at this point that modal operators and quantifiers diverge. Quantifiers can be vacuous. Let φ be a sentence (hence with no free variables). Then $(\forall x)\varphi$ ought to mean nothing more than φ itself. So our final quantificational axiom schema is, simply,

(Q5) $\varphi \supset (\forall x)\varphi$, where φ is a sentence.

When this is added, conventional classical first-order logic is the result.

5. COMPLETENESS, HOW PROVED

In a sense, the proof of the completeness of the quantifier system QC with respect to the model theory presented in the previous section is a "stretching out" of the usual Henkin completeness proof for first-order logic. Let us sketch what we mean by this.

Recall that the usual Henkin argument runs as follows. Take a consistent set, extend it to a ·maximal consistent one, then throw in "witnessess" for the existential quantifiers. But that will destroy maximality, so extend to a maximal consistent set again. But that may add new existential quantifiers, so add new witnesses, and so on. One sequentially alternates a maximal consistent construction with an existential-instantiation construction (and then takes the limit). It will be seen shortly that the completeness proof for QC really amounts to a separation of these two constructions.

First we say what consistency means. A set S is *inconsistent* if, for some finite subset $\{A_1, \ldots, A_n\} \subseteq S$, $(A_1 \wedge \ldots \wedge A_n) \supset \perp$ is a theorem of QC. A set S is *consistent* if it is not inconsistent. We adopt this definition in part because the deduction theorem is not available for QC. Indeed, we do not get the deduction theorem for any of the proper subsystems of quantificational logic presented in section 4.

Next, partition the set of parameters into countably many disjoint sets, each with countably many members: $P_1, P_2, P_3, \ldots$

Let F be a sentence not provable in QC; we produce a countermodel for F. Without loss of generality we assume any parameters of F are in P_1.

First, extend $\{\sim F\}$ to a maximal consistent subset of the set of all sentences with parameters in P_1; call this Γ_F. And set $\mathscr{P}(\Gamma_F) = P_1$.

Next if there are no γ sentences in Γ_F, do nothing (but call Γ_F "queer"). Likewise if there are no δ sentences in Γ_F. do nothing. Otherwise, using rule (Q), it is not hard to show that if $\gamma_1, \gamma_2, \gamma_3, \ldots$ and δ are in Γ_F. and $a \in P_2$ (hence is new to Γ_F) that $\gamma_1(a), \gamma_2(a), \gamma_3(a), \ldots, \delta(a)$ is consistent. (Using Q1 it can be shown that any finite number of γ sentences is equivalent to a single one; this is needed here). This in turn can be extended to a maximal consistent subset of the set of all sentences with parameters in $P_1 \cup \{a\}$. Call it Γ_a. And set $\mathscr{P}(\Gamma_a) = \{a\}$, and $\Gamma_F \mathscr{R} \Gamma_a$. Do this sort of thing for each δ sentence in Γ_F, producing a whole batch of sets $\Gamma_a, \Gamma_b, \Gamma_c, \ldots$, all accessible from Γ_F under $\mathscr{R}$. Now repeat the process with each of these new maximal consistent sets, then with the sets that will arise from that, and so on.

Finally, define $\Vdash$ between the various sets Γ that arise this way, and sentences, as follows. If Γ is "queer", $\Gamma \Vdash \delta$ for all δ. If A is atomic, $\Gamma \Vdash A$ iff $A \in \Gamma$. And finally, the conditions of section 4 can be turned around to say how to define $\Vdash$ (inductively on formula degree) now that we have covered the atomic case. When this is done, a model results, for which $X \in \Gamma \Rightarrow \Gamma \Vdash X$ (but not conversely). And it will be a countermodel to F, since $\sim F \in \Gamma_F$.

Notice that in the proof just sketched, the maximal consistent construction gives the worlds, while the existential instantiation moves things from one world to another. This is what we meant by "stretching out" the Henkin construction.

Adding axioms (Q2)–(Q4) modifies the construction in obvious ways. We leave this to the reader.

Now we can ask, What is the effect on this construction of also imposing that final axiom schema (Q5), $\varphi \supset (\forall x)\varphi$? Very simply, it makes things cumulative. Notice that, in our model, if $\Gamma \mathcal{R} \Delta$, then moving from Γ to Δ, in effect, drops one quantifier from each sentence. But $\varphi \supset (\forall x)\varphi$ allows us to add one quantifier, so the effect is neutralized. Briefly, if we assume (Q5), then if $\Gamma \Vdash \varphi$ and $\Gamma \mathcal{R} \Delta$, then $\Delta \Vdash \varphi$. For, if $\Gamma \Vdash \varphi$, since also $\Gamma \Vdash \varphi \supset (\forall x)\varphi$, we must have $\Gamma \Vdash (\forall x)\varphi$. And since $\Gamma \mathcal{R} \Delta$, then $\Delta \Vdash$ some-instance-of-φ. But since the quantifier was vacuous, this means $\Delta \Vdash \varphi$. Now, that things are cumulative if (Q5) is imposed means the limit (=chain-union) part of Henkin's proof can be carried out. And thus a conventional classical model results.

This sketch must suffice. Details, though slightly devious, are far from devastating.

1. SEMANTIC TABLEAUX

We show how the tableau system of Smullyan [7] for propositional logic may be extended to handle the logics discussed in section 4. We begin with a brief sketch of the system that suffices for propositional logic.

First, proofs are in tree form (written branching downward). There are two *branch-extension rules*:

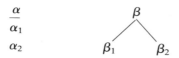

(If α occurs on a branch, α_1 and α_2 may be added to the end of the branch. If β occurs on a branch, the end of the branch may be split, and β_1 added to the end of one fork, β_2 to the end of the other.)

A branch is called *closed* if it contains A and $\sim A$ for some formula A, or if it contains $\bot$, or if it contains $\sim\top$. A tree is called closed if every branch is closed. A closed tree with $\sim X$ at the origin is, by definition, a *proof* of X.

We begin by adding to the above a tableau rule to give the modal logic C. In words, the rule is as follows. If, on a branch, there are ν-formulas, and there is a π-formula, then that π-formula may be replaced by π_0, all the ν-formulas by the corresponding ν_0-formulas, and all other formulas deleted.

We schematize this as follows. First, if S is a set of formulas, define

$$S\# = \{\nu_0 \mid \nu \in S\}.$$

Then the rule is

$$\frac{S, \pi}{S\#, \pi_0} \quad \text{(provided } S\# \neq \phi\text{)},$$

where this is to be interpreted as follows: if $S \cup \{\pi\}$ is the set of formulas on a branch, it may be replaced by $S\# \cup \{\pi_0\}$ (provided $S\#$ is not empty).

EXAMPLE. We show $\Box X \supset \sim\Diamond \sim X$ is provable using this rule.

The proof begins by putting $\sim(\Box X \supset \sim\Diamond \sim X)$ at the origin, then two α-rule applications produce the following one-branch tree:

$$\sim(\Box X \supset \sim\Diamond \sim X)$$
$$\Box X$$
$$\sim\sim\Diamond \sim X$$
$$\Diamond \sim X.$$

Now take S to consist of the first three formulas, and π to be $\Diamond \sim X$. Then $S\# = \{X\}$, which is not empty, so the rule says the set of formulas on the branch may be replaced by $S\# \cup \{\pi_0\}$, namely

$$X$$
$$\sim X$$

and this is closed!

REMARK. Because of the way trees are written, an occurrence of a formula may be common to several branches, but we may wish to modify (or delete) it using the above rule on only one branch. Then, simply, first add new occurrences of the formula at the ends of all the branches that are not to be modified, then use the above rule on the branch to be worked on.

Now the other modal logics can be dealt with easily.

The logic K was axiomatized by adding $\Box\top$. In effect this says there are always ν-formulas available, hence $S\#$ can always be considered to be nonempty. And, in fact, the appropriate tableau system for K is the one above without the provision that $S\#$ be nonempty.

The logic D had $\Diamond\top$ as an axiom. In effect, this says there is always a π-formula around, so an explicit occurrence of π need not be present to apply the tableau rule. Properly speaking, a tableau system for D results from that of K by adding the additional rule:

$$\frac{S}{S\#}.$$

Finally the logic T had as an axiom schema $\nu \supset \nu_0$. Well, simply add the tableau rule

$$\frac{\nu}{\nu_0}$$

(it can easily be shown that the D-rule above is redundant).

For quantifiers we proceed analogously of course. Thus, for a parameter a, and a set S of first-order sentences, let

$$S(a) = \{\gamma(a)\,|\,\gamma \in S\}.$$

Then a tableau system for QC is the Smullyan propositional system plus the rule (to be read in a similar fashion to the one for C above)

$$\frac{S, \delta}{S(a), \delta(a)}$$

provided

(1) a is new to the branch;
(2) $S(a) \neq \phi$.

For QK, drop the requirement that $S(a)$ be nonempty.
For QD, add the rule

$$\frac{S}{S(a)}$$

provided a is new to the branch.

For QT, add the rule

$$\frac{\gamma}{\gamma(a)}$$

for any parameter a.

(This makes the QD rule redundant).

And finally, for first-order logic proper we want the "cumulativeness" that $\varphi \supset (\forall x)\varphi$ brought; that is, as we go on in a tableau construction, sentences should never be deleted. Well, we could replace the QK rule by

$$\frac{S, \delta}{S, S(a), \delta(a)}$$

provided a is new to the branch.

The only other quantifier rule now is

$$\frac{\gamma}{\gamma(a)}.$$

It is not hard to see that these are equivalent to the simpler set

$$\frac{\delta}{\delta(a)} \quad \text{a new} \qquad \frac{\gamma}{\gamma(a)},$$

and we have exactly the first-order system of Smullyan [7].

BIBLIOGRAPHY

[1] Fitting, M.: 1973, 'Model Existence Theorems for Modal and Intuitionistic Logics', *Journal of Symbolic Logic* **38** 613–627.

[2] Kripke, S.: 1963, 'Semantical Analysis of Modal Logic I, Normal Propositional Calculi', *Zeitschrift fur mathematische Logik und Grundlagen der Mathematik* **9**, 67–96.

[3] Kripke, S.: 1965, 'Semantical Analysis of Modal Logic II, Non-normal Modal Propositional Calculi', *The Theory of Models*, J. W. Addison, L. Henkin and A. Tarski (eds.), North-Holland Publishing Co., Amsterdam, pp. 206–220.

[4] Lemmon, E.: 1957, 'New Foundations for Lewis Modal Systems', *Journal of Symbolic Logic* **22** 176–186.

[5] Segerberg, K.: 1971, *An Essay in Classical Modal Logic*, Uppsala.

[6] Smullyan, R.: 1963, "A Unifying Principle in Quantification Theory", *Proceedings of the National Academy of Sciences*.

[7] Smullyan, R.: 1968, *First Order Logic*, Springer-Verlag, Berlin.

Dept. of Mathematics
Lehman College
Bronx, NY 10468
U.S.A.

NICOLAS D. GOODMAN

THE KNOWING MATHEMATICIAN

1.

Mathematics is at the beginning of a new foundational crisis. Twenty years ago there was a firm consensus that mathematics is set theory and that set theory is Zermelo-Fraenkel set theory (ZF). That consensus is breaking down. It is breaking down for two quite different reasons. One of these is a turning away from the excesses of the tendency toward abstraction in post-war mathematics. Many mathematicians feel that the power of the method of abstraction and generalization has, for the time being, exhausted itself. We have done about as much as can be done now by these means, and it is time to return once more to hard work on particular examples. (For this point of view see Mac Lane [16, pp. 37–38].) Another reason for this turning back from abstraction is the economic fact that society is less prepared now than it was twenty years ago to support abstract intellectual activity pursued for its own sake. Those who support research are asking more searching questions than formerly about the utility of the results that can reasonably be expected from projects proposed. Mathematicians, moreover, are increasingly obliged to seek employment not in departments emphasizing pure mathematics, but in departments of computer science or statistics, or even in industry. Thus there is a heightened interest in applied and applicable mathematics and an increased tendency to reject as abstract nonsense what our teachers considered an intellectually satisfying level of generality. The set-theoretic account of the foundations of mathematics, however, is inextricably linked with just this tendency to abstraction for its own sake. Mathematics, on that account, is about abstract structures which, at best, may happen to be isomorphic to structures found in the physical world, but which are themselves most definitely not in the physical world. Thus as mathematicians turn away from pure abstraction, they also become increasingly dissatisfied with the doctrine that mathematics is set theory and nothing else.

There is also another reason for the breakdown of what we may call the set-theoretic consensus on the foundations of mathematics. That is

Synthese **60** (1984) 21–38.   0039–7857/84/0601–0021 $01.80
© 1984 *by D. Reidel Publishing Company*

the breakdown of the consensus within set theory. The work of Gödel and Cohen and Solovay and the rest has shown that Zermelo-Fraenkel is an astonishingly weak theory, which settles few of the issues of central concern to the set theorist. Not only does it not settle the continuum problem, but it also does not settle the Souslin hypothesis, the Kurepa hypothesis, the structure of the analytic hierarchy, or the gap 2 conjecture. A contemporary set theorist, faced with a deep-looking problem, asks first not for a proof or a counterexample, but for a model. He does not expect to prove or refute a conjecture, but to prove it independent. The number of independent set-theoretic axioms grows alarmingly. More and more exotic large cardinals are invented and studied, though none of them can be proved to exist. More and more complex combinatorial principles are extracted from the structure of Gödel's constructible universe or from the technique of some intricate forcing argument and are then shown to hold in some models but not in others. If mathematics is set theory, which set theory is it? No one knows how to choose among these many conflicting principles. This situation, moreover, has gone beyond the point where it is of interest only to logicians. Algebraists interested in the structure of infinite abelian groups must watch their set theory (see [10] and [24]). Analysts interested in the structure of topological algebras must do the same (see [27]). Increasingly it seems that every mathematician whose interests are at all abstract is going to be faced with the problem of choosing which set-theoretic axioms to work with. But it also seems clear that there is no way within the present framework to distinguish which of these alternative set theories is the true one – if that question even makes sense any longer. Each mathematician must rely on his own, increasingly bewildered, intuition or taste. Evidently it is time to try to find a new framework.

2.

I suggest that both of the above difficulties with the set-theoretical foundational consensus arise from the same source – namely, its strongly reductionistic tendency. Most mathematical objects, as they originally present themselves to us, are not sets. A natural number is not a transitive set linearly ordered by the membership relation. An ordered pair is not a doubleton of a singleton and a doubleton. A function is not a set of ordered pairs. A real number is not an equivalence class of

Cauchy sequences of rational numbers. Points in space are not ordered triples of real numbers, and lines and planes are not sets of ordered triples of real numbers. Probability is not a normalized countably additive set function. A sentence is not a natural number. A proof is not a sequence of finite strings of symbols formed in accordance with the rules of some formal system. Each reader can supply his own examples of cases in which mathematical objects have been replaced in our thought and in our teaching by other, purely conceptual, objects. These conceptual objects may form structures which are isomorphic to relevant aspects of the structures formed by the objects we were originally interested in. They are, however, distinct from the objects we were originally interested in. Moreover, they do not fit smoothly into the larger structures to which the original mathematical objects belong.

For example, consider the doctrine that Euclidean three-dimensional space is the set of all ordered triples of real numbers. (See Kaplan [11, p. 70], Randolph [19, p. 69], or Rosenlicht [21, p. 34].) If that doctrine is literally true, then how is it that physical objects are in three-dimensional space? How is it that the physicists assure us there is no preferred coordinate system in physical space? Of course, we can give an account. We may talk about the introduction of a coordinate system, for example. That account, however, would make it clear that the doctrine is not literally true, but is a metaphor or an abuse of language. The student who encounters this doctrine in a course in advanced calculus may not understand that what he is learning is a useful theory about physical space.

For another example, consider the doctrine that a proof is a string of finite sequences of symbols. (See Kleene [13, p. 33] or Enderton [6, p. 546].) If that doctrine is literally true, then to remember a proof must be to remember a string of finite sequences of symbols. Everyone who has ever remembered a complicated proof knows that not to be the case. The student who encounters this doctrine in a course in symbolic logic may not understand that what he is learning is an illuminating theory about correct inference.

In general, the result of a set-theoretic reduction is an object that suffices as a substitute for the original mathematical object in the context of pure mathematics, strictly so-called, but which cannot function as a substitute for the original object outside of pure mathematics. Set-theoretic reductionism leads to a radical oversimplification of the mathematician's world, which makes it appear that that world is disjoint from the universe of discourse of the rest of science. In thinking about pure mathematics we end up no longer thinking about a part of science, but rather about a beautiful, austere substitute for science. This development has meant a progressive impoverishment of the mathematician's intuition. Many mathematicians who know some complex analysis do not know, or do not think of, the connection between analytic functions and the flow of a fluid. When I was taught complex analysis in graduate school, I was not taught that connection. Hence when thinking about analytic functions these mathematicians cannot rely on their commonsense insight into the behavior of water. Again, many mathematicians who know some of the theory of rings and ideals do not know the algorithms worked out by Kronecker and others to compute such ideals. When I was taught ring theory in graduate school, I was not taught those algorithms. For these mathematicians an ideal is not a computational object but merely an abstract set satisfying certain closure conditions. Thus they find it difficult even to consider nontrivial examples of the abstract theory they have learned. Any intuition they may have must be purely formal.

The set-theoretic reductionists have explained away the objects we were trying to study, and now those objects are no longer there to guide us. Perhaps it is time to bring some of them back.

3.

Throughout most of the twentieth century physicists have held that it is not possible adequately to describe the physical world without taking into account the observer who collects the data that the physical theory is intended to predict and explain. Both the theory of relativity and quantum mechanics are in very large part theories of the relation between the observer and the physical reality that he observes. Neither theory, however, denies the reality or the objectivity of the external reality that is being observed. Although both theories involve a far-reaching admixture of epistemology in their basic description of physical reality, neither theory can legitimately be accused of idealism. The observer and the external physical reality with which he is confronted are both irreducibly presupposed. In this sense these theories are fundamentally dualistic.

Twentieth-century mathematicians, on the other hand, have sought to maintain the monistic character of their discipline. The classical mathematician, who is in practice a mathematical Platonist, tries as far as possible to ignore the epistemic aspects of what he does. Although he may make side remarks about heuristic issues or about the degree to which his arguments are effective or constructive, his official pronouncements – his theorems – are austerely free of any reference to the knowing mathematician. Indeed, the orthodox view that mathematics is to be thought of as formalized in a set theory precludes the possibility that mathematical theory contain any reference to the mathematician who knows the theory. There are, of course, set-theoretic versions of some aspects of the epistemology of mathematics. Both recursive-function theory and model theory are set-theoretic reductions of epistemological ideas analogous to the analytic treatment of space. They are quite adequate within certain purely mathematical contexts, but they do not give an analysis in terms of the framework in which these epistemological notions naturally arise. They abstract from any reference to a knower. In this respect classical mathematics resembles classical Newtonian physics. It presupposes a single omniscient knower who plays no role in the theory itself. Alternatively, it is a theory to be thought of as true but not as known.

Constructive mathematicians, on the other hand, have emphasized the role of the knowing mathematician to the exclusion of the reality known. In orthodox intuitionism, for example, it is held that the mathematical objects known are constructed by the knower and have no properties except those known. Mathematical objects exist only insofar as they are known. This is, of course, the basis for the intuitionist's famous rejection of the law of the excluded middle. Thus Brouwer's philosophy of mathematics is a subjective idealism. It denies the existence of any reality external to the knower. In an important sense, mathematics is not knowledge for Brouwer, since mathematical assertions have no truth-value independently of the mathematician's knowledge. For Brouwer, moreover, it is unclear in what sense, if any, mathematics is objective. Thus intuitionistic mathematics is not analogous to contemporary physics. It is more like a hypothetical phenomenalistic account of our experience, which would attempt to predict and explain our sense experiences without assuming the existence of any external reality to which those experiences refer or of which those experiences are experiences. It is now widely held that such a phenomenalistic account is impossible. Similarly, most mathematicians hold that intuitionism cannot suffice as a foundation for any very extensive or meaningful mathematics. Nevertheless, the constructive tradition has the very great merit of having emphasized and studied the epistemic aspects of mathematics. Those aspects have been largely ignored by classical mathematicians.

The debate between the classical and intuitionistic positions on the foundations of mathematics can thus be viewed as a disagreement between two opposed, mutually exclusive, monisms. Each denies that the other's reality is of any fundamental significance. It is like a debate on the foundations of physics between a strict Newtonian who denies that observability is of any fundamental consequence and a strict phenomenalist who denies that our sense experiences refer to any knowable reality outside of ourselves. Obviously such a debate is unlikely to be fruitful. In physics the fruitful step was the step to a dualistic view which emphasized both the role of the observer and the role of the reality being observed. It is the thesis of the present essay that a similar dualistic view on the foundations of mathematics is both possible and desirable.

4.

The observer in quantum mechanics or in the theory of relativity is a very highly idealized physicist. He has no subjective bias. He does not forget anything. He has no deficiencies of experimental technique. His attention never wavers. He never sleeps. In our theory, then, we may expect that the knowing mathematician will be similarly idealized. In particular, we will assume that his powers of concentration are potentially infinite. There is to be no finite bound on the complexity of the computations he can carry out or on the length of the proofs he can construct. As I have argued in another place (see [9]), this is a very considerable idealization of what is the case for human mathematicians or even for the human race viewed collectively as a single mathematician. Thus the knowing mathematician of our theory is himself to be conceived as a mathematical abstraction. His introduction into the theory is a step not toward a constructivistic impoverishment of mathematics, motivated by doubt about the metaphysical underpinnings of set theory, but rather toward an enrichment of classical mathematics by the introduction of a new and more extensive vocabulary. The point is not to consider only those objects which can be known, but rather to consider what aspects of arbitrary objects are knowable in principle.

Thus, although our knowing mathematician is to be thought of as only potentially infinite, he is not to be thought of as somehow situated in the physical world. All he does is mathematics. He has no properties that do not follow from his being an idealized mathematician. In this respect he resembles the idealized physicist of quantum mechanics or of the theory of relativity.

Although our goal is to reverse the set-theoretic reductionism of the last few decades, it is clear that we cannot do that in one fell swoop. It makes no sense to write down a theory having as its primitives all the notions which, in an unanalyzed form, have ever played a role in mathematical practice. Such a theory would be ugly and, presumably, a conservative extension of its set-theoretic fragment. It would give no new insight. The point is not to reintroduce old notions for the sake of not explaining them away. The point, rather, is to try to rebuild our mathematical intuition by gradually enriching it with notions and principles that are not known to be reducible to set theory. Thus I suggest that our first draft of such a theory should be a set theory, but a set theory enriched with intensional epistemic notions. The first theory of the sort I have in mind was based on arithmetic and is to be found in Shapiro [23]. Then Myhill in [17] proposed a theory based on set theory but in which the arithmetic part was still, so to speak, singled out in the very syntax of the theory. Finally, in [8], I proposed a theory that is strictly set theoretic. A similarly motivated but independent and formally very different theory can be found in the recent work of Lifschitz (see [14] and [15]).

The theory I will discuss here is the theory of my [8]. It is a modal set theory, with membership as its only nonlogical primitive, and with Lewis's S4 as its underlying logic. The modal operator of S4 is to be read epistemically. Thus $\Box \phi$ means that ϕ is knowable.

5.

In one respect, at least, our theory will resemble quantum mechanics more than it does the theory of relativity. The whole point of the theory of relativity is to relate the observations of different observers. In the mathematical case this does not seem to make much sense. Troelstra in [26] considers two mathematicians, one of whom communicates numbers to the other without giving him complete information about how those numbers are obtained. The other mathematician, then, is able to make certain predictions about the future behavior of the first mathematician. This entire picture, however, seems to me more closely related to what happens in empirical science than to anything that goes on in mathematics. Mathematicians exchange complete information. They do not generally hide all or part of their algorithms from each other. Moreover, in contrast to the situation in physics, I do not see that it makes sense to suppose that one mathematician somehow has preferential access to some part of mathematical reality. On this basis, then, it seems to me adequate to consider the case of only one knower. This is analogous to the situation in quantum mechanics, where one does not usually consider the effect of having more than one experimenter.

Thus when, in our theory, we write $\square \phi$, we take this to mean that the formula ϕ is knowable to our idealized mathematician. We say "knowable" rather than "known" because of the logic of S4. Let us recall the rules of that system. In addition to the usual rules of the classical predicate calculus, we have the following modal rules:

1. $\quad \square \phi \rightarrow \phi.$

2. $\quad \square \phi \rightarrow \square \square \phi.$

3. $\quad \square \phi \wedge \square(\phi \rightarrow \psi) \rightarrow \square \psi.$

4. $\quad$ From $\vdash \phi$ infer $\square \phi.$

If we take $\square$ to mean 'known', then the third of these rules asserts that whenever our mathematician can make an inference, he has already done so. Even an idealized mathematician presumably does not follow out every possible chain of inference. Thus we must take $\square$ to mean 'knowable'. On this reading, the first of our rules asserts that everything knowable is true. The second asserts that anything knowable can be known to be knowable. One might argue for this by noting that if ϕ is ever known, it will then be known to be known, and hence known to be knowable. The third rule asserts that if both a conditional and its antecedent are knowable, then the consequent is knowable. Finally, the fourth rule derives from our confidence in our very system of axioms. If we actually prove a claim, then we know that claim to be true, and so that claim is knowable.

It seems to have been Gödel in [7] who first observed that S4 can be taken as a formalization of abstract knowability or provability.

6.

I have said that the concept of knowability which our theory formalizes must be taken as a mathematical abstraction. Do we know that that abstraction is coherent? That is to say, can we give an interpretation of our theory that gives some more or less relevant precise account of the modal operator □? The answer to this question was provided by Myhill in [17]. He showed that in his theory the modal operator can be interpreted as provability in that theory itself. Later, in [8], I carried out a similar construction for my theory. From a formal point of view, this is a very satisfactory result. The theory is intended as a formalization of a certain body of mathematical practice, and knowability is formally interpreted as provability in that very formalization. Everything hangs together correctly.

Philosophically, however, it seems very implausible to identify mathematical knowability with provability in some particular formal system. Why should we believe that we can today write down the fundamental principles that will guide all future mathematicians? No past generation of mathematicians was able to write down axioms sufficient for the practice of all future mathematicians. I suggest that this is another example of a set-theoretic reduction. The concept of mathematical knowability in principle is, it seems to me, a fairly clear one. Certainly it is as clear as the concept of an abstract set. That it is clear enough to support axiomatization, and even formalization, is shown by the fact that it has been formalized in S4 in a way that we can all agree is correct, at least as far as it goes. That it is a widespread intuitive notion is shown by the fact, pointed out both by Shapiro and by Myhill, that classical mathematicians in fact use the concept when, for example, they talk of providing an explicit solution of a problem rather than just proving a solution to exist. Whether this concept of knowability is sufficiently clear to support an extensive body of mathematical practice, of course, remains to be seen. We must try to do mathematics explicitly using the notion and find out what difficulties arise. At any rate, to replace this clear and widely shared mathematical intuition by a formal notion like provability in this particular theory is to deprive ourselves of the possible insight to be gained by using the original intuition to guide our mathematical practice.

That there is really something to be gained here can be seen, I think, by looking at the tradition of constructive mathematics. The constructive mathematician is not at all doing the same thing that I am proposing to do. He does not consider himself to be learning about an externally existing reality. Rather, he is studying a reality that he somehow creates by the act of studying it. Nevertheless, epistemic notions play a dominant role in the practice of constructive mathematics. To ask whether an object with certain properties can be constructed is, at least in part, to ask whether an explicit description of an object can be provided in such a way that that explicit description can be known to be a description of an object having those properties. The recent development of constructive analysis by the school of Errett Bishop (see Bishop [1], Bishop and Cheng [2], and Bridges [4]) shows that it is possible to do a great deal of mathematics with such epistemic notions in the forefront of one's attention. Moreover, as has often been pointed out, this work in constructive analysis has been valuable for classical analysis. In the process of constructivizing classical theorems, the constructive analysts have found that they need more explicit information than was needed for the original development. This additional information, then, has led to sharper theorems. In the case of constructive analysis, this fruitful impact of epistemic considerations on classical analysis is hampered by the generally shared conviction that constructive analysis is somehow opposed to classical analysis. Our proposal to add the epistemic notions directly to classical mathematics, without throwing any of classical mathematics away, avoids this problem. In this way we can hope that the additional intuitive richness provided by the epistemic concepts will directly aid in the development of classical mathematics.

7.

The idea of an intuitive mathematical notion more fundamental than any possible set-theoretic definition of the notion may seem strange. We have been trained in accordance with a set-theoretically based standard of rigor according to which a notion that has not been given a set-theoretic analysis is necessarily vague and mathematically unreliable. Everyone feels better when some perfectly clear informal concept is explained as referring to an ordered seven-tuple, the first component of which is a nonempty set and the second component of which is a

function such that . . . . In order to see that there really are mathematical notions independent of set theory, let us consider a familiar example
that is actually rather closely connected with our idea of mathematical
knowability in principle. I am thinking of Church's thesis.

The standard account is that there was a vague and unanalyzed
informal notion of computability in principle, and that Church, Turing,
Kleene, and Post offered alternative analyses of this notion in set-
theoretic terms that turned out to be equivalent. As usual, let us refer to
a function satisfying the formal definition as *recursive*. Church's thesis is
then the vague and premathematical claim that recursiveness coincides
with computability in principle. It has even been urged that the
replacement of computability in principle by recursiveness is analogous
to the replacement of the informal eighteenth-century notion of continuity by the formal nineteenth-century concept defined using epsilons
and deltas (see Shapiro [22]). Note, however, that the two analyses
function quite differently in practice. The informal notion of continuity
is used only heuristically to motivate the epsilon-delta definition. Once
the epsilon-delta definition has been given, no further reference is made
to the informal notion. The practicing analyst who uses the notion of
continuity thinks in terms of the epsilon-delta definition, or in terms of
some other definition equivalent to that definition on the real line.
Every assertion about continuity is explicitly justified by means of the
epsilon-delta definition. The contemporary recursive-function theorist,
on the other hand, uses the informal notion of computability constantly.
He thinks in terms of that notion, rather than in terms of one of the
standard formal definitions. Proofs in the theory of recursive functions
usually no longer refer to the formal definitions at all. If it becomes
relevant to relate the theory to one of the formal definitions, this
connection is established by a global appeal to Church's thesis. The
analyst says, "This function is continuous because I have shown that it
satisfies the epsilon-delta definition." The recursion theorist says, "This
function is recursive because I have shown how to compute it." The
role of the formal, set-theoretic definition in the two cases could not be
more different.

It might be suggested that we are still so close to the time when the
definition of recursiveness was first given that there has not been time
for the formal notion to drive out the informal notion it is intended to
replace. The historical evidence, however, points in the opposite
direction. The early papers in the theory of recursive functions were

very formal and relied directly on some formal definition. Increasingly, however, the literature has moved away from this heavy reliance on what is perceived as a cumbersome and largely irrelevant machine. Recent works avoid the formal definitions altogether. (For early formal expositions, see Kleene [12] or Davis [5]. For a transitional work, see Rogers [20]. He gives the definition in terms of Turing machines, but then never uses it. For an example of the current tendency to avoid the formal definitions entirely, see Shoenfield [25].)

I submit that the situation here is that there was an informal, somewhat vague, notion of computability in principle. The formal work of Kleene and the others served above all to sharpen that informal notion and to give mathematicians the confidence to work with it despite its non-set-theoretic character. Now the informal notion is coming into its own and driving out its formal counterparts except in those situations, such as word problems, where the formal definitions are themselves directly relevant to the problem being studied.

Another example of the same sort, but one that is perhaps so familiar as to be banal, is the concept of logical implication. The formalization of the predicate calculus and the proofs of the soundness and completeness of that calculus provide an analysis of the notion of logical implication. In this case I have no doubt at all that the analysis is correct. Nevertheless, neither I nor any other practicing mathematician checks his inferences by seeing whether they can be formalized in the predicate calculus. The informal and intuitive notion of logical implication, of correct first-order reasoning, is far more fundamental and natural than any analysis we might hope to offer. Here too, I might add, there is an idealization going on. Not every inference which is valid in first-order predicate calculus is feasible for a human being to make. Some first-order formulas are so long that no one will ever write them down. Thus the predicate calculus is an analysis not of actual correct inferability, but of correct inferability in principle.

The example of first-order logic is particularly interesting for our purposes precisely because the analysis is so evidently correct. Not only do we have a definition that is adequate for mathematical purposes, we apparently have a definition that is extensionally correct for all purposes. Nevertheless, that correct set-theoretic account cannot replace the informal notion of which it is an account. The notion remains intuitive, informal, and non-set-theoretic.

8.

Let us concede, for the sake of argument, that there is an intuitive notion of mathematical knowability in principle that is sufficiently clear and sufficiently rich to play a legitimate role in mathematical practice. Nevertheless, there would seem to be a fundamental difficulty about the way in which I am proposing to formalize the notion. Specifically, a modal theory, like the theory of [8], is inherently an intensional theory. Thus as Quine, for example, has emphasized, there may be serious difficulties about the interpretation of such a theory (see Quine [18]). This difficulty comes very much to the fore in Myhill's formulation in [17]. Myhill does not even have full substitutivity of identity. The reason for this difficulty, of course, is that the same set may appear under different descriptions, so that we may know that the set having description A has property P, but not know that the set having description B has the property P, even though, as a matter of fact, these are the same set. In the context of a formal set theory of the sort we are discussing, a set is always specified by giving a defining criterion. Thus when a particular set is specified, we may always suppose that it is specified as the set of all x such that $\phi(x)$, for some formula ϕ. Then the problem comes down to the fact that it may be the case that every x is such that $\phi(x)$ if and only if $\psi(x)$, but it is not knowable that that is the case.

This problem would not arise if everything true were knowable in principle. In that case two specifiable and equal sets would be known to be equal, in principle. But in that case, knowability in principle would coincide with truth, and we would be adding nothing to classical mathematics. The formula $\Box\phi$ would be equivalent to ϕ. But even if it is the case that everything true is knowable in principle, and I find that very implausible, surely we do not know that everything true is knowable. Indeed, it seems likely to me that the set of truths that are knowable in principle and that can be expressed in any single fixed set-theoretic language form a recursively enumerable set. To know that that is not the case is to know that human epistemic potentialities exceed those of any computing machine, no matter what its memory capacity. It seems likely that to know it is not the case would actually be to know that human capacities exceed those of any possible physical mechanism. But that is just not the sort of thing that I think we could ever conceivably know. It is well known, on the other hand, that no recursively enumerable set can exhaust the truths of any nontrivial mathematical theory. Thus, at least as far as we could know, the problem of intensionality does arise.

9.

When we do set-theoretic mathematics, we reason about sets, not about descriptions of sets. Nevertheless, when we reason epistemically about particular sets, asking ourselves questions such as whether we could actually construct a set with a certain property, we are necessarily dealing not with the sets themselves, but only with defining criteria. As a mathematical abstraction, at least, it is not difficult to construct a language in which every set has a defining criterion. For, form a "language" in the logician's sense having a name for every set. Then the set A is defined by the criterion, 'x belongs to A'. Of course, in this language, the set A will also have many other defining criteria, some of which may not be knowably equivalent to this one. Although no human being could learn this language, we often talk as though we were using a finite fragment of it. A mathematician who has proved the existence of a set with a certain property but does not know any actual criterion for membership in such a set will not hesitate to introduce a name for a particular such set. For example, he may write as follows: "Thus we see that there exists a regular ultrafilter. Let D be such an ultrafilter. We know that D has such and such property. Hence . . . ." As an illustration of this procedure, let L be a language of the sort described.

Every set has a defining criterion expressible in L. These criteria can be thought of as representatives of the sets. For all mathematical purposes, these representatives will do the work of the sets. Membership in a set is just satisfaction of the corresponding membership criteria. The usual axiom of extensionality, which asserts that two sets with the same members are identical, tells us that two defining criteria which are satisfied by the same objects are representatives of the same set. Thus, in the spirit of the usual set-theoretic reductionism, we may think of a set as an equivalence class of these defining criteria under the relation of being satisfied by the same objects. (Actually, as this construction is carried out in my paper [8], it is technically somewhat more complex. The problem is that some criteria will not be extensional. Thus sets should be thought of as equivalence classes of hereditarily extensional criteria. But these details need not concern us here.)

The relation of extensional identity is not the only interesting equivalence relation on the defining criteria expressible in our language *L*. A somewhat stronger relation is the relation which holds between two of these criteria when it is knowable that they apply to exactly the same objects. Let me refer to this relation as *modal* extensional identity. Criteria that are modally extensionally identical can be thought of as representatives of the same set-theoretic property or attribute. More formally, by a *property* let us mean an equivalence class of defining criteria under the relation of modal extensional identity. Then we can interpret our modal set theory as being about properties.

In this way, interpreting our theory as being a theory of properties rather than a theory of sets, the problems of interpretation we mentioned above disappear. For, two properties that can be known to apply to the same objects will have exactly the same properties expressible in the language of our theory. Thus if we adopt a modal axiom of extensionality asserting that properties that can be known to apply to the same things are identical, then we will have full substitutivity of identity.

10.

Let me now suggest that the above construction of properties from sets is exactly backwards. The historical development, at least, went in the other direction. Sets first appeared in mathematics in the form of properties expressed in some language – say mathematical German. That is to say, the defining criteria are historically the primary objects. These properties were studied by impredicative and nonconstructive methods, so that it was clear very early that mathematical German was not adequate to formulate all possible such properties (for what amounts to this point, see Borel [3, pp. 109–110]). Thus sets were introduced as mathematical abstractions to support a notion of property that had come detached from the idea of expressibility in any particular, or even any possible, language. The problem then arose about what criterion of identity one should use for these strange new mathematical objects. In the positivistic intellectual climate of the turn of the century, it was not difficult to arrive at a consensus that extensional identity was the only possible criterion. As a matter of fact, however, some sort of intensional identity may be more suitable for pre-set-theoretic mathematical practice. Thus I suggest that we should go back to the situation in, say, 1880, when it was not yet clear what it should mean to say that two sets are identical. In that situation it makes sense to suggest that two sets are identical just in case *it can be known that* they have exactly the same elements. Given that we have an adequate logic of knowability, which we do, this suggestion may be more fruitful than the generally accepted one. To avoid confusion today, however, we should probably not refer to the resulting objects as sets, but as properties.

11.

Once it is clear that the theory we want to write down is a theory of properties in the above sense, it is not difficult to write down the axioms by imitating ZFC (that is, Zermelo-Fraenkel set theory with the axiom of choice). For the details, I refer the reader to [8]. The resulting theory is as rich as set theory. ZFC is faithfully interpretable in it. As a matter of fact, the theory is richer than set theory. For example, in it we can express that we have actually constructed an object, rather than merely proved it to exist. Most of the basic constructive notions are easily expressible. Of course, the underlying metaphysical position is not at all intuitionistic, but Platonistic.

What remains is to try to develop set theory and analysis in this new framework, trying to exploit its additional expressive power. As we do so, we should be able to develop sufficient intuition for the epistemic component of the new framework that we can begin to think informally about the knowing mathematician without having to rely on any formalization, including ours. Then we should be led to ask new questions and to find new phenomena that will enrich classical mathematics. It seems rational to hope that the resulting sharpening of our set-theoretic intuition will either lead to new insight into the apparently unanswerable questions of set theory, or else will enable us to see that those questions are not really very important or very central after all. What is more important is that by working with intensional notions directly connected with our actual mathematical experience – by considering mathematical knowledge, computability, and construction as ingredients of our universe of discourse rather than as merely psychological aspects of our work – we may restore some of the concrete mathematical intuition that we have lost. In this way we may gradually break down the prestige of set-theoretic abstraction and make it possible to bring other concrete aspects of mathematical experience back into the official domain of mathematics. Thus the reintroduction of intensional epistemic notions into mathematical practice may serve as a first step toward the reintegration of mathematics with the rest of science.

Mathematics today seems empty because we have explained away its proper objects of study. We cannot bring those objects back merely by wishing they were still there. Instead we must find mathematical objects that are clearly not sets, which can serve as the forerunners of a new mathematical concreteness. For this role I propose properties, intensionally conceived, and the knowing mathematician.

REFERENCES

[1] Bishop, E.: 1967, *Foundations of Constructive Analysis*, McGraw-Hill, New York.
[2] Bishop, E. and H. Cheng: 1972, *Constructive Measure Theory*, Memoirs Amer. Math. Soc. 116.
[3] Borel, E.: 1898, *Leçons sur la théorie des fonctions*, Gauthiers-Villars, Paris.
[4] Bridges, D. S.: 1979, *Constructive Functional Analysis*, Research Notes in Math. 28, Pitman, London.
[5] Davis, M.: 1958, *Computability and Unsolvability*, McGraw-Hill, New York.
[6] Enderton, H. B.: 1977, 'Elements of Recursion Theory', in J. Barwise (ed.), *Handbook of Mathematical Logic*, North-Holland, Amsterdam, pp. 527–566.
[7] Gödel, K.: 1969, 'An Interpretation of the Intuitionistic Sentential Logic', in J. Hintikka (ed.), *The Philosophy of Mathematics*, Oxford U.P., London, pp. 128–129.
[8] Goodman, N. D.: 1984, 'A Genuinely Intensional Set Theory', in S. Shapiro (ed.), *Intensional Mathematics*, North-Holland Pub. Co., New York.
[9] Goodman, N. D.: 1983, 'Reflections on Bishop's Philosophy of Mathematics', in F. Richman (ed.), *Constructive Mathematics*, Springer Lecture Notes No. 873, pp. 135–145, Reprinted in *The Mathematical Intelligencer* 5 (1983), 61–68.
[10] Hiller, H. L., M. Huber, and S. Shelah: 1978, 'The Structure of Ext(A, Z) and $V = L$', *Math. Zeitschrift* 162, 39–50.
[11] Kaplan, W.: 1973, *Advanced Calculus*, 2nd ed., Addison-Wesley, Reading, Mass..
[12] Kleene, S. C.: 1952, *Introduction to Metamathematics*, Van Nostrand, Princeton.
[13] Kleene, S. C.: 1967, *Mathematical Logic*, Wiley, New York.
[14] Lifschitz, V.: 'On Computable Natural Numbers', Typescript.
[15] Lifschitz, V.: 'The Constructive Component of Classical Mathematics I', Typescript.
[16] MacLane, S.: 1975, 'A History of Abstract Algebra: Origin, Rise, and Decline ot a Movement', in *American Mathematical Heritage, Algebra and Applied Mathematics*, El Paso, Texas, pp. 3–35.
[17] Myhill, J.: 1984, 'Intensional Set Theory', to appear in S. Shapiro (ed.), *Intensional Mathematics*, North-Holland Pub. Co., New York.

[18] Quine, W. V.: 1976, 'Three Grades of Modal Involvement', in *The Ways of Paradox and Other Essays*, rev. ed., Harvard U.P., Cambridge, Mass., pp. 158–176.
[19] Randolph, J. F.: 1968, *Basic Real and Abstract Analysis*, Academic Press, New York.
[20] Rogers, H.: 1967, *Theory of Recursive Functions and Effective Computability*, McGraw-Hill, New York.
[21] Rosenlicht, M.: 1968, *Introduction to Analysis*, Scott, Foresman, Glenview, Ill.
[22] Shapiro, S.: 'On Church's Thesis', Typescript.
[23] Shapiro, S.: 1984, 'Epistemic Arithmetic and Intuitionistic Arithmetic', to appear in S. Shapiro (ed.), *Intensional Mathematics*, North-Holland Pub. Co., New York.
[24] Shelah, S.: 1979, 'On Well Ordering and More on Whitehead's Problem', Abstract 79T-E47, *Notices Amer. Math. Soc.*, 26, A-442.
[25] Shoenfield, J. R.: 1972, *Degrees of Unsolvability*, North-Holland, Amsterdam.
[26] Troelstra, A. S.: 1968, 'The Theory of Choice Sequences', in B. Van Rootselaar and J. F. Staal (eds.), *Logic, Methodology and Philosophy of Science III*, North-Holland, Amsterdam, pp. 201–223.
[27] Williamson, J. H.: 1979, Review of *Topological Algebras* by E. Beckenstein, L. Narici, and C. Suffel, *Bull. Amer. Math. Soc.* (New Series) 1, 237–244.

Dept. of Mathematics
Suny–Buffalo
Amherst, NY 14260
U.S.A.

RAYMOND D. GUMB

"CONSERVATIVE" KRIPKE CLOSURES*

0. INTRODUCTION

Computable Kripke closures are properties of relations which have closures in, roughly speaking, the sense of the transitive closure. They were introduced in [8] to generalize Kripke-style tableaux constructions and were studied from a model-theoretic perspective by Weaver and Gumb [19].

In section 1 of this paper, we review the properties of computable Kripke closures. In section 2, we state four additional laws that can be imposed on computable Kripke closures and state properties of the closures determined by these laws. In a sense, all four of our laws require closures to be "conservative". However, regarding later sections, it is more revealing to classify two of the laws as being commutative and two as being conservative.

In the remaining sections, we sketch applications of these laws in modal logics having a Kripke-style relational semantics: simplifying Kripke-style tableaux constructions (section 3), proving the Craig Interpolation Lemma (section 4), establishing Henkin-style completeness proofs (section 5), and providing a somewhat plausible probabilistic semantics (section 6). At least in modal logic, our laws carve out natural classes of properties of binary relations.

1. COMPUTABLE KRIPKE CLOSURES

The presentation of the computable Kripke closures in this section is much the same as that in [8, section 5]. A model-theoretic study of the first-order Kripke closures using a different notation can be found in [19].

A (*binary*) *relational system* is a pair $\bar{\sigma} = (\sigma, r)$, where σ is a nonvoid set and $r \subseteq \sigma X \sigma$ is a (binary) relation on σ. Let BR be the class of relational systems. Understand $\Pr \subseteq BR$ to be a *property of relations* if Pr is closed under isomorphisms. Let Pr be a property of relations, and let

Synthese **60** (1984) 39–49.   0039–7857/84/0601–0039 $01.10
© 1984 *by D. Reidel Publishing Company*

$\bar{\sigma} = (\sigma, r)$ and $\bar{\sigma}^+ = (\sigma^+, r^+)$ be relational systems. We say that r^+ is a Pr-*relation* (*on* σ^+) if $\bar{\sigma}^+ \in$ Pr. We call r^+ the Pr-*closure of r* (*on* σ^+) and write $r^+ = \mathrm{pr}_{\sigma^+}(r)$ if $\sigma \subseteq \sigma^+$, $r \subseteq r^+$, and r^+ is the smallest Pr-relation on σ^+ containing r. Pr has the *closure property* if the Pr-closure of r on σ exists for every $\bar{\sigma} \in$ BR. Pr has the *monotonic closure property* if Pr has the closure property and, if $\sigma \subseteq \sigma^+$ and $r \subseteq r^+$, then $\mathrm{pr}_{\sigma}(r) \subseteq \mathrm{pr}_{\sigma^+}(r^+)$.

Let $\bar{\sigma}_i = (\sigma_i, r_i) \in$ BR $(i \geq 1)$. The sequence $C_{\mathrm{Pr}} = \bar{\sigma}_1, \bar{\sigma}_2, \dots$ is a (*weak subsystem*) Pr-*chain* if, for every $i \geq 1$, $\bar{\sigma}_i \in$ Pr, $\sigma_i \subseteq \sigma_{i+1}$, and $r_i \subseteq r_{i+1}$. Let C_{Pr} be a Pr-chain. Pr is *closed under unions of* (*weak subsystem*) *chains* if $\bar{\sigma}_w = (\bigcup_{i \geq 1} \sigma_i, \bigcup_{i \geq 1} r_i) \in$ Pr. Pr has the *computable Kripke closure property* if (1) Pr has the monotonic closure property, (2) Pr is closed under unions of chains, (3) Pr is closed under the preimages of (strong) homomorphisms, and (4) whether $\bar{\sigma} \in$ Pr is decidable if σ is finite.

Let Pr have the computable Kripke closure property. Proof that Pr satisfies the following laws can be retrieved from [19]:

(*Operation*) $\mathrm{pr}_{\sigma}(r)$ exists and is uniquely defined for any relational system $\bar{\sigma}$;

(*Monotone*) $\mathrm{pr}_{\sigma}(r) \subseteq \mathrm{pr}_{\sigma^+}(r^+)$ for any relational systems $\bar{\sigma}$ and $\bar{\sigma}^+$ such that $\sigma \subseteq \sigma^+$ and $r \subseteq r^+$;

(*Extensive*) $r \subseteq \mathrm{pr}_{\sigma}(r)$ for any relational system $\bar{\sigma}$;

(*Idempotent*) $\mathrm{pr}_{\sigma}(\mathrm{pr}_{\sigma}(r)) = \mathrm{pr}_{\sigma}(r)$ for any relational system $\bar{\sigma}$;

(*Smallest*) $\mathrm{pr}_{\sigma}(r) \subseteq r^+$ for any relational systems $\bar{\sigma}$ and $\bar{\sigma}^+$ such that $\sigma \subseteq \sigma^+$, $r \subseteq r^+$, and $\bar{\sigma}^+ \in$ Pr;

(*Closure System*) (1) $(\sigma, \sigma^2) \in$ Pr for any nonvoid set σ and (2) $(\sigma \cap \sigma^+, r \cap r^+) \in$ Pr for any relational systems $\bar{\sigma}, \bar{\sigma}^+ \in$ Pr such that $\sigma \cap \sigma^+ \neq \emptyset$;

(*Closure under subsystems*) $(\sigma, r^+ \cap \sigma^2) \in$ Pr for any relational system $\bar{\sigma}^+ \in$ Pr and any nonvoid set σ such that $\sigma \subseteq \sigma^+$;

(*Closure under Preimages of Homomorphisms*) $\bar{\sigma}^+ \in$ Pr if, for some relational system $\bar{\sigma} \in$ Pr and some function j mapping σ^+ onto σ, wr^+w' if and only if $j(w)rj(w')$ for every $w, w' \in \sigma^+$.

(*Closure under Unions of Chains*) Pr is closed under unions of (weak subsystem) chains;

(*Computable*) Whether $\bar{\sigma} \in$ Pr is decidable if σ is finite.

A property of relations is *first-order* if it is the class of models of a set of sentences in the first-order language (with equality) of binary-relational systems, which has a binary-predicate constant and no individual or function constants. A *monotonic universal Horn sentence* is a first-order universal sentence having a matrix either of the form (1) A_0 or (2) $A_1 \& \ldots \& A_n \supset A_0$, where the A_i ($0 \leq i \leq n$) are atomic formulas other than identities. Proof of the following preservation theorem can be found in [19]: a first-order property of relation Pr has the computable Kripke closure property if and only if it is the class of models of a set of monotonic universal Horn sentences and whether $\sigma \in$ Pr is decidable if σ is finite.

The key fact about properties of relations having the computable Kripke closure property (*computable Kripke closures*, for short) is that Kripke tableaux constructions as generalized in the forest method, the deductive system for evolving theories, can be applied whenever the restriction on the accessiblity relation is a computable Kripke closure [8, sections 6–8]. Examples in [8, notes 12 and 14, pp. 93–94] show that the forest method (unless modified) cannot be applied if the restriction on the accessibility relation is not a computable Kripke closure.[1]

2. COMMUTATIVE AND CONSERVATIVE LAWS

Let Pr have the computable Kripke closure property, let $\sigma_0 \neq \emptyset$, let $r_0 \subseteq \sigma_0^2$, let $n \geq 0$, let $1 \leq i \leq n$, let $W_i \notin \sigma_{i-1}$, let $\sigma_{W_i} = \{W_j : 1 \leq j \leq i\}$, let $\sigma_i = \sigma_0 \cup \sigma_{W_i}$, let $W'_{j-1} \in \sigma_{i-1}$, let $r_{W_i} = \{(W'_{j-1}, W_j); 1 \leq j \leq i\}$, and let $r_i = r_0 \cup r_{W_i}$. We state some interesting laws which can be imposed on Pr:[2]

$$(Commutative) \quad \mathrm{pr}_{\sigma_0}(r_1 \cap \sigma_0^2) = \mathrm{pr}_{\sigma_1}(r_1) \cap \sigma_0^2$$
$$= \mathrm{pr}_{\sigma_1}[r_0 \cup \{(W_1, W'_0)\}] \cap \sigma_0^2;$$
$$(Weak\ Commutative) \quad \mathrm{pr}_{\sigma_0}(r_1 \cap \sigma_0^2) = \mathrm{pr}_{\sigma_1}(r_1) \cap \sigma_0^2;$$
$$(Conservative) \quad \mathrm{pr}_{\sigma_0}(r_0) = \mathrm{pr}_{\sigma_1}(r_1) \cap (\sigma_1 X \sigma_0);$$
$$(Weak\ Conservative) \quad \mathrm{pr}_{\sigma_n}(r_n) \cap (\sigma_{W_n} X \sigma_0) = \emptyset.$$

It is readily demonstrated that law (*Commutative*) is equivalent to the law dubbed (*Unsafe Conservative*) in [8, p. 92] and that law (*Conservative*) as stated here is equivalent to its formulation in [9].[3]

We write (K) for the class of computable Kripke closures, *Symmetric* for the member of K such that $\bar{\sigma} \in Symmetric$ iff r is a symmetric relation on σ, (*Commutative*) for the subclass of (K) having members satisfying law (*Commutative*), etc. (*Nonconservative*) is the subclass of

(K) having members which do not satisfy law (*Weak Conservative*). We have:

PROPOSITION 1.
   (a) (*Commutative*) $\subset$ (*Weak Commutative*),
   (b) (*Conservative*) $\subset$ (*Weak Conservative*),
   (c) (*Conservative* = (*Weak Commutative*) $\cap$ (Weak Conservative),
and
   (d) (*Commutative*) $\cap$ (*Nonconservative*) $\neq \emptyset$.

Proof (a) Obviously, (*Commutative*) $\subseteq$ (*Weak Commutative*). Since the property of relations determined by the sentence $\forall x_1 x_2 (x_1 R x_2 \supset x_2 R x_2)$ belongs to (*Weak Commutative*) but not to (*Commutative*), the inclusion is proper. (b) By proposition 1 in [9], we have (*Conservative*) $\subset$ (*Weak Conservative*). Since *Quasi-Reflexive* $(\forall x_1 x_2 (x_1 R x_2 \supset x_1 R x_1))$ belongs to (*Weak Conservative*) but not to (*Conservative*), the inclusion is proper. (c) ($\subseteq$) By part (b) and the definitions of σ_1 and r_1, we have (*Conservative*) $\subseteq$ (*Weak Conservative*) and (*Conservative*) $\subseteq$ (*Weak Commutative*). Hence, (*Conservative*) $\subseteq$ (*Weak Commutative*) $\cap$ (*Weak Conservative*). ($\supseteq$) Suppose Pr $\in$ (*Weak Commutative*) $\cap$ (*Weak Conservative*))–(*Conservative*). Since Pr $\in$ (*Weak Commutative*) and Pr $\notin$ (*Conservative*), we have $(w_1, w) \in$ $\text{pr}_{\sigma_1}(r_1)$ for some $w \in \sigma_0$. But $(w_1, w) \notin \text{pr}_{\sigma_1}(r_1)$ by law (*Weak Conservative*). Contradiction. Hence, (*Conservative*) $\supseteq$ (*Weak Commutative*) $\cap$ (*Weak Conservative*). (d) *Symmetric* $\in$ (*Commutative*) $\cap$ (*Nonconservative*). $\qquad\square$

We turn now to delimiting the the first-order computable Kripke closures which satisfy the above laws. Let $n \geq 2$. Understand the following formulas to be universally quantified (i.e., to be the matrices of monotonic universal Horn sentences), and understand the variables $x_i (1 \leq i \leq n)$ to be distinct:

   (a) xRx,
   (b) $x_1 R x_2$,
   (c) $x_1 R x_2 \,\&\, \ldots \,\&\, x_{n-1} R x_n \supset x_1 R x_n$,
   (d) $x_1 R x_2 \,\&\, \ldots \,\&\, x_{n-1} R x_n \supset x_n R x_1$,
   (e) $x_1 R x_2 \,\&\, \ldots \,\&\, x_{n-1} R x_n \supset x_i R x_n \ (1 < i \leq n)$,
   (f) $x_1 R x_2 \,\&\, \ldots \,\&\, x_{n-1} R x_n \supset x_i R x_j \ (1 \leq i \leq j < n)$,
   (g) $x_1 R x_2 \,\&\, \ldots \,\&\, x_{n-1} R x_n \supset x_j R x_i \ (1 < i < j \leq n \geq 3)$, and

(h) $x_1 R x_2 \& \ldots \& x_{n-1} R x_n \supset x_j R x_i \ (1 \le i < j < n \ge 3)$.

The corresponding modal characteristic axioms can be retrieved from the literature. We say that a restriction on the accessibility relation *countenances backward-looping* if it implies a sentence of one of the forms (b), (d), (g), or (h).

The Table I indicates which sentences of the above forms can be used to axiomatize properties of relations belonging to the classes determined by our four laws:

TABLE I

CLASS	AXIOM SCHEMAS. $\sqrt{} = $ Yes							
	(a)	(b)	(c)	(d)	(e)	(f)	(g)	(h)
(*Commutative*)	$\sqrt{}$	$\sqrt{}$	$\sqrt{}$	$\sqrt{}$				
(*Weak Commutative*)	$\sqrt{}$	$\sqrt{}$	$\sqrt{}$	$\sqrt{}$	$\sqrt{}$			
(*Conservative*)	$\sqrt{}$		$\sqrt{}$			$\sqrt{}$		
(*Weak Conservative*)	$\sqrt{}$		$\sqrt{}$			$\sqrt{}$	$\sqrt{}$	

Clearly, all computable Kripke closures which countenance backward-looping are nonconservative, but some nonconservative computable Kripke closures do not have an axiomatization consisting soley of axioms of the forms (a)–(h).[4]

3. KRIPKE-STYLE TABLEAUX CONSTRUCTIONS

In [8, pp. 91–92], we remarked that the forest method can be simplified by having only one overall relation r for all cords of branches when Pr satisfies law (*Commutative*) (called there (*Unsafe Conservative*)). The simplification works because, for r_b the relation on σ_b a cord of branches, we have $r_b = r \cap \sigma_b^2$ by law (*Commutative*).

In [8], the underlying logic of evolving theories is free tense logic with equality. We now consider the modal analogue of evolving theories as in [9]. In the forest method, when we update r_b as a result of an application of D-rule (D. $\sim \square$), we compute $\mathrm{pr}_{\sigma_b}(r_b \cup \{(w, w')\})$ (w' does not belong to the old value of σ_b) [8, pp. 58–59], never $\mathrm{pr}_{\sigma_b}(r_b \cup \{(w', w)\})$. Hence, in the modal analogue of evolving theories, we need only one overall relation r when Pr satisfies law (*Weak Commutative*) because $r_b = r \cap \sigma_b^2$ by law (*Weak Commutative*). So, by proposition 1(a), for the modal analogue of evolving theories, the simplified version of the forest method is correct (i.e., sound) for a broader class of Kripke closures than it is for evolving theories proper.[5]

4. THE CRAIG INTERPOLATION LEMMA

In the remaining sections, unless explicitly stated otherwise, we consider free modal logics (and their tense analogues) in which neither the Barcan formula $\forall x \Box A \supset \Box \forall x A$ nor its converse $\Box \forall x A \supset \forall x \Box A$ nor the formula $T \neq T' \supset \Box T \neq T'$ holds (as in [1, 8, 9]).

From results of Fine [3], it can be seen that the Craig Interpolation Lemma fails in (1) free tense logics in which the restriction on the accessibility relation is any computable Kripke closure [1, p. 208] and (2) free modal logics in which the restriction on the accessibility relation is a computable Kripke closure countenancing backward-looping. As announced in [9], the Craig Lemma does go through in free modal logics with equality in which the restriction on the accessibility relation is a conservative computable Kripke closure, and this positive result extends to the free version of S4 with equality and increasing domains into which **IQC***= ((free) intuitionistic logic with equality [14]) naturally translates [11]. The Craig Lemma also holds in **IQC***= [7].

By proposition 1(c), the simplification of the forest method discussed in the preceding section works when Pr ∈ (*Conservative*). So, given two theories which are jointly unsatisfiable, the simplification of the forest method somewhat reduces the complexity of constructing a separating sentence in [9].

5. HENKIN-STYLE COMPLETENESS PROOFS

Let $\mathscr{L}$ be a free modal (tense) logic having (1) a computable Kripke closure $Pr_{\mathscr{L}}$ as the restriction on the accessibility relation and (2) a Henkin-style completeness proof as adapted by Makinson [17] for its sentential fragment. To obtain a Henkin-style completeness proof for $\mathscr{L}$, the trick is – roughly speaking – to insure that the individual parameters at world w are included in those at w' if we must have wRw' for R to be a $Pr_{\mathscr{L}}$-relation. So, Leblanc's construction [13] using Makinson attendants for the free modal versions of **K**, **M**, and **S4** go through for $\mathscr{L}$ (with only minor editing to show that R is a $\mathrm{Pr}_{\mathscr{L}}$-relation) whenever $\mathscr{L}$ is a modal logic and $\mathrm{Pr}_{\mathscr{L}} \in$ (*Weak Conservative*). In the related construction of Barnes and Gumb [1] for $\mathbf{QK_t^*}=$, the minimal free tense logic with equality, Makinson attendants are also used. However, the construction is supplemented with a "pull-back" phase: if $HA\,(GA)$ is in a future (past) attendant of a point in time t, then A is pulled-back into t. The Barnes-Gumb construction goes through (with minor editing) whenever (1) $\mathscr{L}$ is a tense logic or (2) $\mathscr{L}$ is a modal logic and $\mathrm{Pr}_{\mathscr{L}} \subseteq$ *Symmetric*.[6] In each of these cases, all truth-value assignments in the Barnes-Gumb construction have the same domain. I conjecture that the construction can be extended to any $\mathscr{L}$ such that $\mathrm{Pr}_{\mathscr{L}}$ countenances backward-looping. However, the editing can be quite complex in these systems since it is possible that, for some distinct w and w', $\sim wR^n w'$ for any $n \geq 1$, and so not all truth-value assignments can have the same domain. In any case, for the more complicated modal and mixed modal-tense systems, Fine's method of diagrams [4], which does not use Makinson attendants, recommends itself.

6. PROBABILISTIC SEMANTICS

As I mentioned in an earlier talk before this Association [6], probabilistic semantics can be extended to certain other logics. As many of these results can now be found in the literature,[7] I will confine my remarks to a point bearing on the importance of computable Kripke closures countenancing backward-looping.

Let $\mathscr{L}$ and $\mathrm{Pr}_{\mathscr{L}}$ be as in the preceding section. For the moment, we restrict our attention to $\mathscr{L}$ a free tense logic. Let Σ be the natural numbers or a nonempty initial segment of them, Let $M = (I, P, \Sigma, R, J, f, d, \alpha)$ be a $\mathrm{Pr}_{\mathscr{L}}$-metaphor [8], let $A, B \in L(J, P)$, let $i, j \in \Sigma$, and let P_i and P_j be Popper-Carnap probability functions for $L(J, P)$ [6] satisfying the following two additional conditions:

$$(B11_G)\ P_i(GA, B) = \operatorname*{Limit}_{n \to \infty} \overset{m}{\underset{j=1}{X}} P_{i,j}(GA, B),$$

where $m = $ minimum $(n, \operatorname{card}(\Sigma))$ and

$$\begin{aligned} P_{i,j}(GA, B) &= P_i(A, B) \text{ if } i = j \ \&\ iRi, \\ &= P_j(A, HB) \text{ if } i \neq j \ \&\ iRj, \text{ and} \\ &= 1 \text{ otherwise;} \end{aligned}$$

and

$$(B11_H)$$

which is defined in the obvious (mirror-image) manner.

Let Prob $= \{P_i : i \in \Sigma\}$. A $\mathrm{Pr}_{\mathscr{L}}$ -*probability structure* is a 7-tuple of the form $M_P = (I, P, \Sigma, R, J, f, \mathrm{Prob})$.

By the Löwenheim-Skolem-Kripke theorem [8, p. 83], a $\mathrm{Pr}_{\mathscr{L}}$-satisfiable theory S has a $\mathrm{Pr}_{\mathscr{L}}$-metaphor having a countable time domain Σ. Given adequate conditions for identity and the free quantifiers,[8] conditions $(B11_G)$ and $(B11_H)$ assure the completeness of $\mathscr{L}$ with respect to probabilistic semantics.

To see this, consider the Barnes-Gumb completeness proof for $\mathbf{QK_t}^* =$ (edited for metaphor theory). Take $v_i(A)$ $(A \in L(J,P), i \in \Sigma)$ to be 1 if A is true on Asgn (M, i) and to be 0 otherwise. Define $P_i(A,B)$ the *Popper-Carnap correlate of* v_i to be $v_i(B \supset A)$ for every A, $B \in L(J,P)$ [6]. So if S is consistent, then S is $\mathrm{Pr}_{\mathscr{L}}$-satisfiable in the probabilistic sense (i.e., for every $B, C \in S$, $\dot{f}(B)$ and $\dot{f}(C)$ are P_0-normal, and $P_0(f(C), f(B)) = 1$, where f is any one-one function such that $f : I \to J$ and card $(J\text{-Image }(f)) = \aleph_0$).

It is readily verified that the axiom of necessitation (if A is an axiom, then so are GA and HA) is valid for all $\mathscr{L}$ and that the characteristic axiom for reflexive systems $(GA \supset A$ and its mirror image) is valid for any $\mathscr{L}$ such that $\mathrm{Pr}_{\mathscr{L}} \subseteq$ *Reflexive*. Some additional plausible condition(s), however, seems to be required for the validity of, e.g., the axiom of distribution $(G(A \supset B) \supset (GA \supset GB))$ and the characteristic axioms for tense systems $(A \supset GPA$ and its mirror image).

I take it that condition $(B11_G)$ would be less plausible if we took $P_{i,j}(GA, B)$ to be, e.g., $P_j(A, \text{true})$ rather than $P_j(A, HB)$ when i $\neq$ j and iRj in defining $P_i(GA, B)$. The time j lies in the future of time i, and so the condition HB is relevant at time j because B is relevant at time i. However, I suspect that some practitioners of probabilistic semantics will not agree with me on this point: One man's implausible is another man's probable.

A probabilistic semantics for $\mathscr{L}$ is readily extended to $\mathrm{Pr}_{\mathscr{L}}$-evolving theories having amenable counterparts [8, p. 82]. Further, a similarly plausible condition $(B11_\square)$ for modal $\mathscr{L}$ can be given whenever the Barnes-Gumb construction goes through (with minor editing) as described in the preceding section. (To obtain $(B11_\square)$ from $(B11_G)$ read '$\square$' for both 'G' and 'H'.)

NOTES

* This paper summarizes work performed under a 1978-80 grant from the Foundation for the Advancement of Interdisciplinary Studies for research on evolving theories, intensional logics, probabilistic semantics, and applications of them to the semantics of natural languages. I have been interested in applications of logic to the semantics of natural languages for some time [10, p. 104]; needless to say, applications of the systems discussed in this paper stand in need for further articulation. I would like to express my appreciation to my friends Bob Barnes, Hugues Leblanc, and George Weaver for having worked with me over the past few years.

[1] An example in [19] shows that the forest method cannot be applied to the free version (with equality) of a logic considered by Lemmon [16, p. 74]. The case for the free version of Boolos's **G** is more complicated. Even if we replace our condition $(\underline{D}. \sim \Box)$ on model systems [9] with Boolos's [2, p. 113]

$(\underline{D}. \sim \Box_{\mathbf{G}})$ If $\sim \Box A \in s(w)$, then, for some $w' \in \sigma$ such that wrw', $\sim A, \Box A \in s(w')$.

we do not have completeness. Let Pr be the property of being a transitive, well-capped relation. Then the sentence

$$\sim \Box p \,\&\, \Box \exists x \forall y \exists z \sim \Box Lyz$$

is not Pr-satisfiable but the forest method will never pronounce it to be Pr-inconsistent. This incompleteness result *for the forest method* is to be expected because Pr does not satisfy law (*Closure under Unions of Chains*). This does not show, of course, that the theorems of free **G** are not recursively enumerable or even that no extension of the tree method can work for it.

[2] We could also define law:

(*Kuratowski*) $\mathrm{pr}_{\sigma_n}(r_n) = \mathrm{pr}_{\sigma_n}(r_0) \cup \mathrm{pr}_{\sigma_n}(r_{w_n})$, and $\mathrm{pr}_{\sigma_0}(\emptyset) = \emptyset$

to induce a topology on the class of binary-relational systems. The first-order Kuratowski computable Kripke closures include those having a set of axioms of the form $x_1 R x_2 \supset x_3 R x_4$, where the x_i $(1 \le i \le 4)$ need not be distinct individual variables. The class of properties of binary relations determined by law (*Kuratowski*) overlaps the classes determined by the four laws in the text and is included in none.

However, I know of no application of law (*Kuratowski*) in intensional logics.

[3] The first equivalence is obvious. For the second equivalence, it suffices to show that $\mathrm{pr}_{\sigma_1}(r_1) = \mathrm{pr}_{\sigma_1}(\mathrm{pr}_{\sigma_0}(r_0) \cup \{(W'_0, W_1)\})$. Since $r_0 \subseteq \mathrm{pr}_{\sigma_0}(r_0)$ by law (*Extensive*) of computable Kripke closures, we have the $\subseteq$-inclusion by law (*Monotone*). Since $\mathrm{pr}_{\sigma_1}(r_1) = \mathrm{pr}_{\sigma_1}(\mathrm{pr}_{\sigma_1}(r_1))$ by law (*Idempotent*), $(W'_0, W_1) \in \mathrm{pr}_{\sigma_1}(r_1)$ by law (*Extensive*), and $\mathrm{pr}_{\sigma_0}(r_0) \subseteq \mathrm{pr}_{\sigma_1}(r_1)$ by law (*Monotone*), we have the $\supseteq$-inclusion.

[4] For modal characteristic axioms, see for example [18]. We conjecture that no Pr $\in$ (*Nonconservative*)-(*Backward-Looping*) can be the restriction on the accessibility relation in a modal logic (having a modal characteristic axiom). Fine [8, pp. 49–50 and note 7, pp. 90–91] has shown that there is no modal characteristic axiom correspoiding to the restriction expressed by $\forall xy(xRx \supset xRy)$.

[5] If evolving theories are outfitted with function symbols, condition $(\underline{D}. =)$ on model systems [8, p. 37] must be strengthened to read:

($\underline{D.} =$)$_{func}$ If $A \in s(t)$ is a literal and $T = T' \in s(t)$, then $A(T'/T)$, $A(T/T') \in s(t)$,

for Hintikka's Theorem [8, Theorem 4.1, p. 42] would not go through otherwise and we would not have completeness. To see this, consider the unsatisfiable theory (in free logic with equality) $\{f(T) \neq T, T = f(T)\}$. If, as in condition ($\underline{D.}=$), we did not require $A(T/T') \in s(t)$ (but only $A(T'/T) \in s(t)$), in the forest method we would never derive a sentence of the form $T' \neq T'$ violating C-rule ($\underline{C.} \neq$).

[6] Roughly, the Barnes-Gumb construction [1, pp. 196–99] is edited as follows. Read '$\Box$' for 'G' and 'H', read '$\mathcal{N}$' for '$\mathcal{F}$' and '$\mathcal{P}$', etc. Some clauses become redundant. So, for example, we have r' is a Makinson attendant of r iff $\mathbf{I}(r') = r$ and $\mathbf{T}(r, j) = r'$ for some j such that $A_j \in \mathcal{N}$. The construction can be simplified somewhat using devices of Gabbay [5, pp. 195–202].

[7] When I wrote [6], I did not know of Harper's pioneering work on probabilistic semantics [12] and so did not credit him for his outstanding achievements.

[8] These can be found in Harper [12]. Alternatively, take the conditions B9–B10 for equality in [6], replace condition B7 with B7*, and add condition B11:

B7*.
$$\mathbf{Pr}(\forall xA, B) = \underset{n \to \infty}{\text{Limit}} \overset{n}{\underset{i=1}{X}} Q_i, \text{ where}$$
$$Q_i = \mathbf{Pr}(A(T_i/x), B) \text{ if } \mathbf{Pr}(\exists x(x = T_i), B) = 1$$
$$= 1 \text{ otherwise.}$$

B11.
$$\mathbf{Pr}(T = T', B) = 0 \quad \text{if } \mathbf{Pr}(\exists x(x = T), B) = 1 \text{ but}$$
$$\mathbf{Pr}(\exists x(x = T'), B) = 0.$$

Regarding completeness, it suffices to show that condition B11 holds when v is as in [6, sections 5 and 6] (adapted to the free case) and $\mathbf{Pr}$ is the Popper-Carnap correlate of v, since the case for B7* is much the same as for B7'. So suppose $\mathbf{Pr}(\exists x(x = T), B) = 1$ but $\mathbf{Pr}(\exists x(x = T'), B) = 0$. Then $v(B) = 1$ because B is $\mathbf{Pr}$-normal. Hence, $v(\exists x(x = T)) = 1$ but $v(\exists x(x = T')) = 0$ and hence $v(\forall x(x \neq T')) = 1$. By the (provable) specification law $\forall xA \supset (\exists x(x = T) \supset A(T/x))$, we have $v(T \neq T') = 1$, so $v(T = T') = 0$. Hence; $\mathbf{Pr}(T = T', B) = 0$. Regarding soundness, it suffices to show that $\mathbf{Pr}(\forall y \exists x(x = y), B) = 1$ for any $\mathbf{Pr}$ and any $\mathbf{Pr}$-normal wff B. By condition B7*, $\mathbf{Pr}(\forall y \exists x(x = y), B) < 1$ if there is some m such that $\mathbf{Pr}(\exists x(x = T_m), B) < 1$ and $\mathbf{Pr}(\exists x(x = T_m), B) = 1$, which is impossible.

I prefer Harper's condition for the free universal quantifier over B7* for the same reason that I prefer Leblanc's condition B7 over B7': Harper's condition does not presume the logical independence of the instances of a universal quantification. More plausible conditions for identity are still needed.

Conditions B7* and B11 were presented in my earlier talk before this Association but were not included in my written paper [6].

REFERENCES

[1] Barnes, R. F. and R. D. Gumb: 1979, 'The Completeness of Presupposition Free Tense Logic', *Zeitschrift für Mathematische Logik und Grundlagen der Mathematik* (hereafter *ZMLGM*) **25**, 193–208.

[2] Boolos, G.: 1979 *The Unprovability of Consistency*. Cambridge: Cambridge University Press.

[3] Fine, K: 1979 'Failures of the Interpolation Lemma in Quantified Modal Logic', *Journal of Symbolic Logic*, **44** 201–206.

[4] Fine, K.: 1978, 'Model Theory for Modal Logic, Part I–The *De Re/De Dicto* Distinction', *Journal of Philosophical Logic*, **7** 125–156.

[5] Gabbay, D.: 1975 'Model Theory for Tense Logics', *Annals of Mathematical Logic* **8**, 185–236.

[6] Gumb, R. D., 'Comments on Probabilistic Semantics', [15].

[7] Gumb, R. D. 'The Craig-Lyndon Interpolation Lemma for (Free) Intuitionistic Logic with Equality', [15].

[8] Gumb, R. D.: 1979, *Evolving Theories*. New York: Haven, 1979.

[9] Gumb, R. D.: 1984, 'An Extended Joint Consistency Theorem for a Family of Free Modal Logics with Equality', *Journal of Symbolic Logic*, **49**.

[10] Gumb, R. D.: 1972, *Rule-Governed Linguistic Behavior*, Mouton, The Hague.

[11] Gumb, R. D. 'A Translation of (Free) Intuitionistic Logic with Equality into Free S4 with Equality and Increasing Domains', [15].

[12] Harper, W. L.: 'A Conditional Belief Semantics for Free Quantificational Logic with Identity', in [15].

[13] Leblanc, H.: 1976, *Truth-Value Semantics* North Holland, Amsterdam.

[14] Leblanc, H. and Gumb, R. D.: 'Completeness and Soundness Proofs for Three Brands of Intuitionistic Logic', in [15].

[15] Leblanc, H., Gumb, R. D., and Stern, R. (eds.): 1983, *Essays in Semantics and Epistemology*, Haven, New York.

[16] Lemmon, E. J.: 1977, *An Introduction to Modal Logic*, Basil Blackwell, Oxford.

[17] Makinson, D. C.: 1966, 'On Some Completeness Theorems in Modal Logic' *ZMLGM* **12** 379–384.

[18] Sahlqvist, H.: 1975, 'Completeness and Correspondence in the First and Second Order Semantics for Modal Logic' in S. Kanger (ed.), *Proceedings of the Third Scandinavian Logic Symposium, Uppsala* 1973, North Holland, Amsterdam.

[19] Weaver, G. and R. D. Gumb' 1982, 'First Order Properties of Relations with the Monotonic Closure Property', *ZMLGM*, **28**, 1–5.

Computer Science Dept.
California State University – Northridge
Northridge, CA 91330
U.S.A.

Added in Proof

Though not noted in [19], Pr is also closed under homomorphisms. As revealed by William Hanson's review of [8] (*Journal of Symbolic Logic* **47** (1982), p. 456), closure under homomorphisms is needed in the proof of Theorem 8.2 [8], a version of the compactness theorem for evolving theories. So we also have the law:

(*Closure under Homomorphisms*) $\bar{\sigma} \in$ Pr if, for some relational system $\bar{\sigma}^+ \in$ Pr and some function j mapping σ^+ onto σ, $j(w)rj(w')$ if and only if wr^+w' for every w, $w' \in \sigma^+$.

Proof: Pr is closed under subsystems and isomorphisms. □

HENRY HIŻ

FREGE, LEŚNIEWSKI AND INFORMATION SEMANTICS ON THE RESOLUTION OF ANTINOMIES

1. INTRODUCTION

Frege sharply distinguished functions from objects. The interplay between the two domains led to serious complications to which many logicians of our century have addressed themselves. Perhaps the time has come to bridge this gap and to this end information semantics is a contribution.

Frege supposed that every function uniquely determines an object which is the value range of the function. He assumed also that functions with the same value range apply to the same objects. As is well known, Frege's postulates led to a contradiction. In order to analyze the problems involved in the antinomial character of Frege's theory, it is advisable to abstract from the intuitive sense of such wordings as 'is the value range of' (or 'is a class of'), and to note the relation by using the arbitrary letter 'α' and to see in it only what is stated in the postulates. Two of Frege's assumptions can then be formulated as follows:

A1. $\wedge f \vee a \ulcorner a \alpha f \urcorner$.

A2. $\wedge afgx \ulcorner a \alpha f \wedge a \alpha g \wedge f(x) . \supset g(x) \urcorner$.[1]

In Frege's way of thinking, to be an object is to be an admissible argument of a function.[2] Syntactically, this amounts to the position that, in the formula '$a \alpha f$', 'a' is of the same grammatical category as any possible (syntactic) argument of 'f'. Then 'a' is available for substitution (under the usual restrictions on substitution) for a variable occurring as a (syntactic) argument to 'f'.[3] Such substitutions are blocked both by Russell's ramified type theory and by the metatheory of Tarski.[4]

In Section 2 of the present paper, the derivation of a contradiction from A1 and A2, in essence, follows Leśniewski's formulation which, in turn, is close to Frege's rendering of Russell's reasoning. In Section 3 and 4, I will report the details of Leśniewski's work. I will abstract from

Synthese **60** (1984) 51–72.   0039–7857/84/0601–0051 $02.20

the less popular features of Leśniewski's theories. I will place these formulations entirely in the second-order predicate logic.

Leśniewski wrote a book about antinomies. It was never published and, so far as I know, the only handwritten copy of it vanished in Warsaw in 1944. Sobociński published an extensive paper reconstructing in detail some of the main ideas and proofs.[5] Sobociński's paper is my main source. Leśniewski has shown that A2 and a weaker form of A1, viz.,

A1′. $\bigwedge fx^\mathsf{r} f(x) \supset \bigvee a^\mathsf{r} a\, \alpha\, f^\mathsf{n}.$

also lead to a contradiction (Section 3). It was proven by Leśniewski, and independently by Quine, that a weakened form of A2, namely, the statement proposed by Frege as an ad hoc repair of his theory, when used with A1′, also leads to a contradiction (Section 4).[6] In Section 5, I recall briefiy the thoughts of Leśniewski that led to the rejection of A2 and to the subsequent developments of both metalogic by Tarski and mereology by Leśniewski.

Information semantics is a theory which studies what is said by sentences. It is derived from the work of Carnap and from a currently accepted linguistic theory which treats sentences as the main vehicle of semantic properties. Sections 6–9 contain the approach of information semantics to the antinomy.[7] Section 6 shows that if f in A1 and A2 is a sentence, i.e., a function from zero arguments, no antinomy arises. Rather, a semantic theorem takes place, a theorem asserting that no sentence says that what it says is false. *Says that* is further examined in Section 7. Using the relation of *saying that* one can define truth (Section 8). The concept of information, as it applies to the present problem, is presented in Section 9.

The concept of a sentence is to be understood here broadly as any phrase which can be a consequence; a is a sentence if and only if $\bigvee X^\mathsf{r} X \neq \phi \wedge \iota a \subset Cn(X)^\mathsf{l}$. Thus, in the language of arithmetic, not only such strings as '$2 + 4 = 17$' are sentences but also such as '$x + y = 17$' and '$\bigvee x^\mathsf{r} f(x)^\mathsf{n}$'. There are sentences which do not say anything. A conjunction of two sentences, each of which says something, may not say anything. Logic, usually, studies the limiting case only, where the conjunction of two assertions forms an assertion. To partly bridge the gap between logic and natural language one has to consider these more complex cases.

In this paper I will disregard the fact that some sentences are ambiguous. The concept of a sentence used here is therefore what sometimes is called 'a reading of a sentence'.[8]

2. RUSSELL'S ANTINOMY

A1. $\bigwedge f \bigvee a ^{\lceil} a \, \alpha \, f ^{\rceil}$

A2. $\bigwedge afgx ^{\lceil} a \, \alpha \, f \wedge a \, \alpha \, g \wedge f(x). \supset g(x) ^{\rceil}$

DR. $\bigwedge a ^{\lceil} R(a) \equiv . \bigvee f ^{\lceil} a \, \alpha \, f ^{\rceil} \wedge \bigwedge f ^{\lceil} a \, \alpha \, f \supset \sim f(a) ^{\rceil \rceil}$

A3. $\bigwedge a ^{\lceil} a \, \alpha \, R \supset \sim R(a) ^{\rceil}$ (DR, f/R)

A4. $\bigwedge af ^{\lceil} a \, \alpha \, R \wedge a \, \alpha \, f. \supset \sim f(a) ^{\rceil}$

Proof. $\bigwedge af$, if

1.	$a \, \alpha \, R$	
2.	$a \, \alpha \, f$, then	
3.	$\bigwedge x ^{\lceil} f(x) \supset R(x) ^{\rceil}$	(A2, 2, 1)
4.	$\sim R(a)$	(A3, 1)
	$\sim f(a)$	(3, 4)

A5. $\bigwedge a ^{\lceil} a \, \alpha \, R \supset R(a) ^{\rceil}$

Proof. $\bigwedge a$, if

1.	$a \, \alpha \, R$, then	
2.	$\bigvee f ^{\lceil} a \, \alpha \, f ^{\rceil}$	(1)
3.	$\bigwedge f ^{\lceil} a \, \alpha \, f \supset \sim f(a) ^{\rceil}$	(A4, 1)
	$R(a)$	(DR, 2, 3)

A6. $\bigwedge a ^{\lceil} \sim a \, \alpha \, R ^{\rceil 19}$ (A3, A5)

3. NO VACUOUS OBJECTS

It may be thought that the antinomy is due to A1, i.e., to the supposition that every property determines an object. If f is a property which does not belong to any object and if '$a \alpha f$' is understood as saying that a is the class (in the collective sense) of objects with the property f, then a must be a vacuous object. The conception of vacuous objects seems fictitious and A1 should be rejected, in that reading. Leśniewski, therefore, instead of A1, accepted:

A1′. $\bigwedge fx \ulcorner f(x) \supset \bigvee a \ulcorner a \alpha f \urcorner \urcorner$.

However, A1′ together with A2 leads to the conclusion that there is at most one object. This by itself does not contradict A1′, A2 or any of their consequences. However, it is unacceptable; for if one believes in the existence of individual objects, then one usually believes in the existence of many individual objects.

Here then is the modified Leśniewski's derivation of the identity of all individuals from A1′ and A2.

A1′. $\bigwedge fx \ulcorner f(x) \supset \bigvee a \ulcorner a \alpha f \urcorner \urcorner$

A7. $\bigwedge a \ulcorner {\sim} R(a) \urcorner$ (A1′, A6)

A8. $\bigwedge af \ulcorner a \alpha f \supset f(a) \urcorner$

   Proof. $\bigwedge af$, if

   1. $a \alpha f$, then

                $\bigvee g$

   2. $a \alpha g$

   3. $g(a)$ (DR, A7, 1)

                $f(a)$ (A2, 3, 1, 2)

Dι. $\bigwedge ab \ulcorner \iota b(a) \equiv a = b \urcorner$

A9. $\bigwedge ab \ulcorner a \alpha \iota b \supset a = b \urcorner$ (A8, $f / \iota b$, D)

A10. $\bigwedge abc \ulcorner a \alpha (\iota b \cup \iota c) \wedge a = c . \supset b = c \urcorner$

   Proof. $\bigwedge abc$, if

   1. $a \alpha (\iota b \cup \iota c)$

   2. $a = c$, then

3. $(\iota b \cup \iota c)(a)$ (A8, 1)

4. $\iota b(a) \vee \iota c(a)$ (3)

if

4.1.1. $\iota b(a)$, then

4.1.2. $\iota b(c)$ (4.1.1, 2)

4.1.3. $b = c$ (4.1.2)

if

4.2.1. $\iota c(a)$, then

        $\vee d$

4.2.2. $d \; \alpha \; \iota c$ (A1', 4.2.1)

4.2.3. $d = c$ (A9, 4.2.2)

4.2.4. $d = a$ (4.2.3, 2)

4.2.5. $a \; \alpha \; \iota c$ (4.2.2, 4.2.4)

4.2.6. $(\iota b \cup \iota c)(b)$

4.2.7. $\iota c(b)$ (A2, 1, 4.2.5, 4.2.6)

4.2.8. $b = c$ (4.2.7)

        $b = c$ (4, 4.1.3, 4.2.8)

A11. $\wedge acb^{\ulcorner} a \; \alpha \; (\iota c \cup \iota b) \wedge a = c \,.\, \supset b = c^{\urcorner}$ (A10)

A12. $\wedge bc^{\ulcorner} b = c^{\urcorner}$

Proof. $\wedge bc$,

1. $(\iota b \cup \iota c)(b)$

        $\vee a$

2. $a \; \alpha \; (\iota b \cup \iota c)$ (A1', 1)

3. $(\iota b \cup \iota c)(a)$ (A8, 2)

4. $\iota b(a) \cup \iota c(a)$ (3)

if

4.1.1. $\iota b(a)$, then

4.1.2. $b = c$ (A11, 2, 4.1.1)

if

4.2.1. $\iota c(a)$, then

4.2.2. $b = c$ (A10, 2, 4.2.1)

         $b = c$ (4, 4.1.2, 4.2.2)

4. FREGE'S AMENDMENT

In view of the antinomy, Frege suggested an amendment to his theory. The amendment consists in adding to A2 a condition that the x in question should not be α-related to f. As is well known, this amendment does not suffice. In 1938 Leśniewski proved that the amended theory leads to the conclusion that there are no more than two objects. Leśniewski's proof was published in the Sobociński's paper cited in Note 5. Quine writes that "Leśniewski's argument, as set forth by Sobociński, is hard to dissociate from special features of Leśniewski's system".[10]

In this section, I will repeat Leśniewski's argument, dissociating it from those "special features of Leśniewski's system" which are difficult to comprehend by people not trained in Warsaw. As in the preceding two sections, the proof is set in the predicate logic of (at least) second order.

Using the notation explained in Note 9, I reformulate the postulates A1', the amended A2, and a postulate, implicit in Frege, that requires that classes defined by the same condition be identical.

B1'. $\bigwedge fx \, {}^\lceil f(x) \supset \bigvee a \, {}^\lceil \alpha \langle f \rangle (a){}^\rceil{}^\rceil$

B2. $\bigwedge fagb \, {}^\lceil \alpha \langle f \rangle (a) \wedge \alpha \langle g \rangle (a) \wedge f(b) \wedge \sim(\alpha \langle f \rangle (b)) \, . \supset g(b){}^\rceil$

B3. $\bigwedge fab \, {}^\lceil \alpha \langle f \rangle (a) \wedge \alpha \langle f \rangle (b) \, . \supset \, = \langle b \rangle (a){}^\rceil$

The derivation of B15:

B4. $\bigwedge fb \, {}^\lceil f(b) \supset \bigvee a \, {}^\lceil \alpha \langle \alpha \langle f \rangle \rangle (a){}^\rceil{}^\rceil$.

    Proof. $\bigwedge fb$, if

1. $f(b)$, then

       $\bigvee a$

2. $\alpha\langle f\rangle(a)$ (B1′, 1)

       $\bigvee a^{\mathsf{r}}\alpha\langle\alpha\langle f\rangle\rangle(a)^{\mathsf{1}}$ (B1′, x/a, $f/\alpha\langle f\rangle$, 2)

I.e., if f is a property that belongs to something, then there is the class of those things which have the property of being the class of things that have the property f.

5. $\bigwedge gbfa^{\mathsf{r}}g(b) \wedge \alpha\langle f\rangle(a) \wedge \alpha\langle g\rangle(a) . \supset . a = b \vee f(b)^{\mathsf{1}}$

Proof. $\bigwedge gbfa$, if

1. $g(b)$

2. $\alpha\langle f\rangle(a)$

3. $\alpha\langle g\rangle(a)$, then

4. $\alpha\langle g\rangle(b) \vee f(b)$ (B2, 2, 3, 1)

       $a = b \vee f(b)$ (4, B3, 3)

B6. $\bigwedge fbcg^{\mathsf{r}}f(b) \wedge \alpha\langle\alpha\langle f\rangle\rangle(c) \wedge \alpha\langle\alpha\langle g\rangle\rangle(c) .$

                         $\supset \bigvee a^{\mathsf{r}}\alpha\langle f\rangle(a) \wedge . c = a \vee \alpha\langle g\rangle(a)^{\mathsf{n}}$

Proof. $\bigwedge fbcg$, if

1. $f(b)$

2. $\alpha\langle\alpha\langle f\rangle\rangle(c)$

3. $\alpha\langle\alpha\langle g\rangle\rangle(c)$, then

       $\bigvee a$

4. $\alpha\langle f\rangle(a)$ (B1′, 1)

5. $c = a \vee \alpha\langle g\rangle(a)$ (B5, $g/\alpha\langle f\rangle$, b/a, a/c, $f/\alpha\langle g\rangle$, 4, 3, 2)

       $\bigvee a^{\mathsf{r}}\alpha\langle f\rangle(a) \wedge . c = a \vee \alpha\langle g\rangle(a)^{\mathsf{1}}$ (4, 5)

B7. $\bigwedge abcg^{\mathsf{r}}g(b) \wedge \alpha\langle\alpha\langle g\rangle\rangle(a) \wedge \alpha\langle g\rangle(c) \wedge a = c . \supset g(a)^{\mathsf{1}}$

Proof. $\bigwedge abcg$, if

1. $g(b)$

2. $\alpha\langle\alpha\langle g\rangle\rangle(a)$

3. $\alpha\langle g\rangle(c)$

4. $a = c$, then

5. $\alpha\langle g\rangle(a)$ $(3,4)$

6. $\alpha\langle g\rangle(b)$ $(B2, f/g, g/\alpha\langle g\rangle, 5, 2, 1)$

7. $a = b$ $(B3, f/g, 5, 6)$

        $g(a)$ $(1, 7)$

B8. $\wedge fbcg \, {}^{\ulcorner} f(b) \wedge \alpha\langle\alpha\langle f\rangle\rangle(c) \wedge \alpha\langle\alpha\langle g\rangle\rangle(c) \wedge g(c) . \supset f(c)^{\urcorner}$

   Proof. $\wedge fbcg$, if

1. $f(b)$

2. $\alpha\langle\alpha\langle f\rangle\rangle(c)$

3. $\alpha\langle\alpha\langle g\rangle\rangle(c)$

4. $g(c)$, then

        $\vee a$

5. $\alpha\langle f\rangle(a)$

6. $c = a \vee \alpha\langle g\rangle(a)$ $(B6, 1, 2, 3)$

7. $c = a \vee f(c)$ $(6, B5, b/c, 4, 5)$

        $f(c)$ $(7, B7, g/f, a/c, c/a, 1, 2, 5)$

D$\bar{R}$. $\wedge a \, {}^{\ulcorner} \bar{R}(a) \equiv . \vee f \, {}^{\ulcorner} \alpha\langle f\rangle(a)^{\urcorner} \wedge \wedge f \, {}^{\ulcorner} \alpha\langle\alpha\langle f\rangle\rangle(a) \supset {\sim} f(a)^{\urcorner\urcorner}$

B9. $\wedge ab \, {}^{\ulcorner} \alpha\langle\alpha\langle \bar{R}\rangle\rangle(a) \supset {\sim} \bar{R}(b)^{\urcorner}$.

   Proof. $\wedge ab$, if

1. $\alpha\langle\alpha\langle \bar{R}\rangle\rangle(a)$, then

2. $\vee f \, {}^{\ulcorner} \alpha\langle f\rangle(a)^{\urcorner}$ (1)

3. ${\sim}\bar{R}(a)$ $(D\bar{R}, a/\bar{R}.1)$

        $\vee f$

4. $\quad\quad \alpha\langle\alpha\langle f\rangle\rangle(a)$

5. $\quad\quad f(a)$ $\quad\quad\quad\quad\quad$ (D$\bar{R}$, 2, 3)

$\quad\quad {\sim}\bar{R}b)$ $\quad\quad\quad\quad\quad\quad$ (B8, 1, 4, 5)

B10. $\wedge b'{\sim}\bar{R}(b)'$ $\quad\quad\quad\quad\quad\quad$ (B4, B9)

B11. $\wedge fb\,'f(b) \supset \vee a\,'\alpha\langle\alpha\langle f\rangle\rangle(a) \wedge f(a)''$

Proof. $\wedge fb$, if

1. $\quad\quad f(b)$, then

$\quad\quad \vee a$

2. $\quad\quad \alpha\langle\alpha\langle f\rangle\rangle(a)$ $\quad\quad\quad\quad$ (B4, 1)

3. $\quad\quad \vee f'\alpha\langle f\rangle(a)'$ $\quad\quad\quad\quad\quad$ (2)

$\quad\quad \vee g$

4. $\quad\quad \alpha\langle\alpha\langle g\rangle\rangle(a))$

5. $\quad\quad g(a)$ $\quad\quad\quad\quad\quad$ (D$\bar{R}$, B10, 3)

6. $\quad\quad f(a)$ $\quad\quad\quad\quad\quad$ (B8, 1, 2, 4, 5)

$\quad\quad \vee a\,'\alpha\langle\alpha\langle f\rangle\rangle(a) \wedge f(a)'$ $\quad\quad\quad\quad$ (2, 6)

B12. $\wedge bde'{\sim}(b = d) \wedge \alpha\langle\alpha\langle\iota b \cup \iota d\rangle\rangle(e) \wedge d = e . \supset \alpha\langle\iota d\rangle(b)'$

Proof. $\wedge bde$, if

1. $\quad\quad {\sim}(b = d)$

2. $\quad\quad \alpha\langle\alpha\langle\iota b \cup \iota d\rangle\rangle(e)$

3. $\quad\quad d = e$, then

4. $\quad\quad (\iota b \cup \iota d)(b)$

5. $\quad\quad \iota d(d)$

$\quad\quad \vee c$

6. $\quad\quad \alpha\langle\alpha\langle\iota d\rangle\rangle(c)$

7. $\quad\quad \iota d(c)$ $\quad\quad\quad\quad\quad$ (B11, 5)

8. $\quad\quad \iota e(c)$ $\quad\quad\quad\quad\quad$ (7, 3)

9.	$\alpha\langle\alpha\langle\iota b \cup \iota d\rangle\rangle(c)$	(2, 8)
	$\vee a$	
10.	$\alpha\langle\iota d\rangle(a)$	
11.	$c = a \vee \alpha\langle\iota b \cup \iota d\rangle(a)$	(B6, 7, 6, 9)
	$\vee x$	
12.	$\alpha\langle\iota b \cup \iota d\rangle(x)$	
13.	$c = x \vee \alpha\langle\iota d\rangle(x)$	(B6, 4, 9, 6)
14.	$c = a \vee a = x$	(11, B3, 12)
15.	$c = x \vee a = x$	(13, B3, 10)
16.	$a = x$	(14, 15)
17.	$\alpha\langle\iota d\rangle(x)$	(10, 16)
18.	$x = b$	(B5, 4, 17, 12, 1)
	$\alpha\langle\iota d\rangle(b)$	(17, 18)

B13. $\wedge bde^{\mathfrak{r}} \sim(b = d) \wedge \alpha\langle\alpha\langle\iota b \cup \iota d\rangle\rangle(e) \wedge d = e \,.\, \supset \alpha\langle\iota b\rangle(d)^{1}$

Proof. $\wedge bde$, if

1.	$\sim(b = d)$	
2.	$\alpha\langle\alpha\langle\iota b \cup \iota d\rangle\rangle(e)$	
3.	$d = e$, then	
4.	$\alpha\langle\iota d\rangle(b)$	(B12, 1, 2, 3)
5.	$\iota b(b)$	
6.	$\iota d(d)$	
	$\vee a$	
7.	$\alpha\langle\alpha\langle\iota b\rangle\rangle(a)$	
8.	$a = b$	(B11, 5)
9.	$\alpha\langle\iota d\rangle(a)$	(4, 8)

10. $\sim(a = d)$ (8, 1)

             $\alpha\langle \iota b\rangle(d)$ (B5, 6, 7, 9, 10)

B14. $\wedge bd^{\mathsf{r}}\sim(b = d) \supset \alpha\langle \iota d\rangle(b)$

Proof. $\wedge bd$, if

1. $\sim(b = d)$, then

2. $(\iota b \cup \iota d)(b)$

             $\vee e$

3. $\alpha\langle\alpha\langle \iota b \cup \iota d\rangle\rangle(e)$

4. $(\iota b \cup \iota d)(e)$ (B11, 2)

5. $\alpha\langle\alpha\langle \iota d \cup \iota b\rangle\rangle(e)$ (3)

6. $e = b \vee e = d$ (4)

             $\alpha\langle \iota d\rangle(b)$ (6, B13, 1, 5, B12, 1, 3)

B15. $\wedge abc^{\mathsf{r}}\sim(a = b) \wedge \sim(c = b) . \supset a = c^{\mathsf{l}}$

Proof. $\wedge abc$, if

1. $\sim(a = b)$

2. $\sim(c = b)$, then

3. $\alpha\langle \iota b\rangle(a)$ (B14, 1)

4. $\alpha\langle \iota b\rangle(c)$ (B14, 2)

             $a = c$ (B3, 3, 4)

If instead of B1′ you use A1, you can conclude that $\wedge ab^{\mathsf{l}}a = b^{\mathsf{l}}$, as is shown in the paper by Sobociński quoted in Note 5, pp. 227–28.

5. LEŚNIEWSKI ON ANTINOMIES

Leśniewski's resolution of the antinomy consists of drawing three distinctions. First, antinomies are of two kinds: some are semantical, others are set-theoretical. If 'α' is 'denotes', 'expresses', 'defines', 'represents', 'says that' or the like, the antinomy is semantical. On the other hand, if 'α' is 'is a class of' or 'is a set given by', then the antinomy is set-theoretical. Second, for the semantical antinomies, when one is speaking about a language, one is not speaking in that language, but in its meta-language. This distinction was further developed by Tarski. The resolution of the semantical antinomies proposed by Leśniewski and Tarski is closely related to the ramified type theory of Russell. (This point is persuasively argued by Church[4].) Third, for the set-theoretical antinomies, the word 'class' can be understood in its distributive sense, i.e., predicative sense, or else it can be understood in its collective sense. Thus, Pennsylvania is a collection of counties. It is also a collection of townships and a collection of mountains, valleys, plateaus, rivers and lakes – or rather parts of them, those parts that are in Pennsylvania. A class, in the collective sense, is an object which is a composite of objects which are its parts. Leśniewski concluded that no logical concept, 'logical' in the sense of Frege, can play the role of a class which Frege designed. The concept of a collective is a useful and reasonable substitute. Leśniewski then formed a theory, Mereology, that deals with the part-whole relation. The applications to mathematics are obvious and some of them are among those that Frege had in mind. The point $\langle 1, -3 \rangle$ is a part of the parabola given by the relation $y = x^2 - 4x$. The parabola is the class of pairs of points in that relation, and it is a class of two half-parabolas: $y = x^2 - 4x$ where $x \le 2$ and $y = x^2 - 4x$ where $x \ge 2$ (with an overlapping point). An example well-known among logicians of an operation that forms a collection out of its parts is concatenation. In the mereological interpretation of '$a \, \alpha \, f$' as 'a *is a collection of objects with the property* f', A2 is wrong again. Because a text is a concatenation of sentences as well as a con-catenation of words.[11]

6. ZERO-ARGUMENT FUNCTIONS

A1 and A2, as well as DR, were intended by Frege, and by Russell, for functions of any number of arguments. What has been said here about functions of one argument can easily be repeated about functions of two arguments, of three arguments, etc. Let us consider functions of zero arguments, even though it is not clear whether Frege ever intended to include this case. A functor of zero arguments is a constant. When it is of the grammatical category of a sentence, it is a sentence, and that is what the structures of A1 and A2 indicate. A sentential function of zero arguments is that which a sentence describes. Sometimes it is called "a proposition". I will call it here "a situation", with the understanding that not all situations are actually occuring; some are actual happenings or facts, others are only possible. Instead of 'f', 'g', etc., of zero arguments it is customary to write 'p', 'q', etc., and this notation will be adopted here. It happens that Frege's principles when applied to functions of no arguments do not lead to a contradiction; Russell's antinomy is not reconstructible in this case. 0A6 below says that no sentence a is α-related to Ra. This does not contradict 0A1; there may be another sentence b which is α-related to Ra. Some interpretations of '$a \, \alpha \, p$' are natural and lead to new approaches to some classical semantical problems.

A1 and A2 became

0A1. $\wedge p \vee a \ulcorner a \, \alpha \, p \urcorner$

0A2. $\wedge apq \ulcorner a \, \alpha \, p \wedge a \, \alpha \, q \wedge p . \supset q \urcorner$

Following the line of reasoning that led to A6 we obtain 0A6.

0DR. $\wedge a \ulcorner R(a) \equiv . \vee p \ulcorner a \, \alpha \, p \urcorner \wedge \wedge p \ulcorner a \, \alpha \, p \supset \sim p \urcorner \urcorner$

0A3. $\wedge a \ulcorner a \, \alpha \, R(a) \supset \sim R(a) \urcorner$ (0DR, $p/R(a)$)

0A4. $\wedge ap \ulcorner a \, \alpha \, R(a) \wedge a \, \alpha \, p . \supset \sim p \urcorner$

Proof. $\wedge ap$, if

1. $a \, \alpha \, R(a)$

2. $a \, \alpha \, p$, then

3. $p \supset R(a)$ (0A2, $q/R(a), 2, 1$)

4. $\sim R(a)$ (0A3, 1)

       $\sim p$ (3, 4)

0A5. $\wedge a \ulcorner a \, \alpha \, R(a) \supset R(a) \urcorner$

Proof. $\wedge a$, if

1. $a \, \alpha \, R(a)$, then

2. $\vee p \ulcorner a \, \alpha \, p \urcorner$ (1)

3. $\bigwedge p^\ulcorner a \,\alpha\, p \supset \sim p^\urcorner$ (0A4, 1)

          $R(a)$ (0DR, 2, 3)

0A6. $\bigwedge a^\ulcorner \sim(a \,\alpha\, R(a))^\urcorner$ (0A3, 0A5)

What remains of Frege's theory in this particular case is provably consistent. Let the variable 'a' range over two particular sentences, one true, the other false, e.g., '$\bigwedge q^\ulcorner q \supset q^\urcorner$' and '$\bigwedge q^\ulcorner q^\urcorner$'. And let '$\alpha$' be interpreted as the usual equivalence relation from the sentential logic. Then 0A1 and 0A2 are true, 'R' is the negation, and 0A6 is a tautology. This proves the consistency of 0A1 with 0A2.

However, 'p' and 'a' in 0A1 and 0A2 may be read as belonging to two different grammatical categories. In the derivation of 0A6 this possibility is respected. The operational aspect of that possibility is that we do not substitute for 'a' any concrete sentence; for 'p', on the other hand, we substitute, e.g., '$R(a)$'. '$R(a)$' is of the sentential grammatical category no matter what the grammatical category of the argument 'a'. The proof of consistency of 0A1 with 0A2 can be modified to fit this case; '$a \,\alpha\, p$' may be interpreted as 'a is the truth-value of p' and 'a' may range over the two truth-values. Again 0A1 and 0A2 are true under this interpretation.

According to the Leśniewski-Tarski theory, 0A6 is in a mixed language; 'a' is used both in a metalanguage (the last occurrence) and in the meta-metalanguage (the last but one occurrence). The substitution of '$R(a)$' for 'p' in the proof of 0A3 is not allowed by their theory, nor by the ramified theory of types of Russell. But in the restricted case considered now there seems to be no harm in relaxing the rigidities of these theories. In a language, we often speak about phrases of that very language. It is only self-referentiality that is the source of complications. 0A6 informs us that no sentence can say that what it says is false. But another sentence of the same language may say that a sentence says that that last sentence is false. It has been claimed, for instance by Zellig Harris, that a self-referential supposed sentence really is not an English sentence.[12] 0A6 justifies that claim; '$a \,\alpha\, R(a)$' either is not a sentence or it does not say what it is supposed to say.

7. SAYING THAT

What was said in Section 6 suggests an interpretation of '$a \,\alpha\, p$' as 'a

says that p'. We will call it Interpretation 1. In this interpretation a may be a sentence in a language, or a longer text composed of sentences. Also it may be a mark in a system of marks, for instance, a road sign or an X-ray of patient's lungs. A red traffic light says that everybody who approaches it had better stop. An X-ray picture may say that the patient has tuberculosis, and the color of wheat that it is ready for harvest. I will write '$S\langle p\rangle(a)$' for 'a says that p', and *mutatis mutandis* for other cases of arguments of 'says that'.

In Interpretation 1, 0A2 becomes S1:

S1. $\wedge paq\ulcorner S\langle p\rangle(a) \wedge S\langle q\rangle(a) \wedge p . \supset q\urcorner$

This is true. But 0A1 in this interpretation is not true. It is not true that for every possible situation p there is a sentence in a given language (or a mark in a system of marks) which says that p. There may be, indeed, there must be indescribable situations – indescribable in a given language. To see this, recall first of all that the expressive capabilities of languages differ; what can be said in one language may not be sayable in another. Secondly, a is understood as a finite mark, for instance, as a finite sequence of sentences. But the situation p may be not adequately describable by a finite set of sentences; the adequate system of sentences may not be finitely axiomatizable and therefore not expressible by a finite conjunction 'a'. Therefore, S2 holds,

S2. $\sim\wedge p\vee a\ulcorner S\langle p\rangle(a)\urcorner$

contrary to 0A1. The same objections hold for a statement which corresponds to A1',

$$\wedge p\ulcorner p \supset \vee a\ulcorner S\langle p\rangle(a)\urcorner$$

Thus, S3 holds:

S3. $\sim\wedge p\vee a\ulcorner p \supset S\langle p\rangle(a)\urcorner$

No matter what the language, there are situations, and there are situations that actually occur, which cannot be expressed in the language.

Similarly, there are syntactically well-formed sentences of a language which do not say anything. One of them is '$a \alpha R(a)$' or, in Interpretation 1, '$S\langle R(a)\rangle(a)$'. Another is 'The first denominator of a triangle is a root of a square of a group'. Or, 'She was acquitted of innocence'. What is and what is not a sentence partly depends on a particular grammatical theory. The vagaries of grammatical theories do not correspond to anything that happens in the world.

To say that a says that p is to say that the consequences of a are the same as the consequences of all sentences which say that p. Thus, S4 holds:

S4. $\bigwedge paX'S\langle p\rangle(a) \supset Cn(\iota a \cup X) = Cn(S\langle p\rangle \cup X)^1$

This formulation of S4 notes that consequences are rarely drawn from a single sentence. Usually one also uses as premises common knowledge, the theorems of the science presupposed for the study (such as mathematics for physics), etc. The concept of consequence used here is that elaborated by Tarski in his address to the 1935 Paris Congress of Scientific Philosophy.[13]

In addition to what I called in previous sections the Leśniewski-Tarski theory, Tarski developed in detail a strong semantics in which the concept of consequence was defined and also for some kinds of language, the concept of truth. In this strong semantical theory Tarski uses the concepts of language, of translation, and of denotation without defining them. The concept of saying is not used, at least not explicitly. I try to avoid the concept of denotation as philosophically doubtful, and those concepts which use denotation in their definitions, such as satisfaction.

Using 'S' one can restate several known semantical theorems, for instance, the Deduction Theorem:

S5. $\bigwedge qpX'S\langle q\rangle \subset Cn(S\langle p\rangle \cup X) \supset S\langle p \supset q\rangle \subset Cn(X)^{1\ 14}$

8. TRUTH

The concept of truth as elaborated by Tarski can be defined using the concept of saying that as follows:

DTr. $\bigwedge a'Tr(a) \equiv \bigvee p'S\langle p\rangle(a) \wedge p^{11}$

A sentence is true just when a situation occurs about which the sentence speaks, or when it is as the sentence says. I think that this definition, or Tr1 below, captures the intent of the Convention T of Tarski (the book quoted in Note 13, 187–88) and it overcomes some of the difficulties mentioned there in connection with the use of a sentence and the use of a name for it. It is why DTr is a definition whereas Convention T was not. It may be, however, that the gain is not as substantial as it seems, because the concept of saying used in DTr is complicated and S1–S5 of Section 7 do not elucidate it sufficiently.

From DTr and S1 it follows that:

Tr1. $\wedge pa^r S\langle p\rangle(a) \supset . \mathrm{Tr}(a) \equiv p^1$

Provided that $S\langle p\rangle(a)$, the proof of the implication '$p \supset \mathrm{Tr}(a)$' is immediate. The proof of the reverse implication's as follows: if

1. $S\langle p\rangle(a)$

2. $\mathrm{Tr}(a)$, then

       $\vee q$

3. $S\langle q\rangle(a)$

4. q (DTr, 2)

       p (S1, 1, 3, 4)

By introducing the definition of falsehood,

DFl. $\wedge a^r \mathrm{Fl}(a) \equiv \vee p^r S\langle p\rangle(a) \wedge \sim p^{11}$

it is easy to prove, by a similar reasoning that led to Tr1:

Tr2. $\wedge pa^r S\langle p\rangle(a) \supset . \mathrm{Fl}(a) \equiv \sim p^1$

From Tr1 and Tr2, it follows that:

Tr3. $\wedge a^r \sim \vee p^r S\langle p\rangle(a)^1 \vee \mathrm{Tr}(a) \vee \mathrm{Fl}(a)^1$

Quartum non datur. Every sentence is either true, or false, or does not say anything. This does not deny the law of excluded middle

$$\bigwedge p^\Gamma p \lor \sim p^\urcorner$$

which, however, does not speak about sentences and their content.

9. INFORMATION

By the information of a sentence a relative to a set (of sentences) X, we understand the set of consequences of the set X augmented by a which are not consequences of X, provided that a says something.

DInf. $\bigwedge aXY^\Gamma \mathrm{Inf}\langle a, X\rangle(Y) \equiv \bigvee p^\Gamma S\langle p\rangle(a) \land Y = \mathrm{Cn}(\iota a \cup X) - \mathrm{Cn}(X)^\urcorner$

In view of S4 from Section 7 one can equally well say DInf':

DInf'. $\bigwedge aXY^\Gamma \mathrm{Inf}\langle a, X\rangle(Y)$
$$\equiv \bigvee p^\Gamma S\langle p\rangle(a) \land Y = \mathrm{Cn}(S\langle p\rangle \cup X) - \mathrm{Cn}(X)^\urcorner$$

The information of a sentence varies with what is assumed. If X is our knowledge prior to the hearing of a, then $\mathrm{Inf}\langle a, X\rangle$ is what we learned from a. If X is a text, or a conversation that preceded the utterance of a, then the information in a is what is conveyed beside what was so far conveyed. In this case the reader does not have to agree with the text. In various sublanguages of a natural language there are different sets of assumptions and therefore a sentence may have different information content depending upon which sublanguage it is used in.[15] This explains some (but only some) kinds of ambiguity.

In Interpretation 2, '$Y \alpha p$' will be

$$\lq \bigvee a^\Gamma S\langle p\rangle(a) \land \mathrm{Inf}\langle a, X\rangle(Y)^\urcorner$$

In other words '$Y \alpha p$' is now 'Y is the information, relative to X, of a sentence saying that p'. It is indeed a set of interpretations, one for each X.

Let us recall a few standard theorems about the concept of consequence:

Cn 1. $\bigwedge X^\Gamma X \subset \mathrm{Cn}(X)^\urcorner$

Cn 2. $\bigwedge XY^\Gamma \mathrm{Cn}(X) \cup \mathrm{Cn}(Y) \subset \mathrm{Cn}(X \cup Y)^\urcorner$

Cn 3. $\bigwedge X^r Cn(Cn(X)) \subset Cn(X)^1$

Cn 4. $\bigwedge X^r X \subset \mathrm{Tr} \supset Cn(X) \subset \mathrm{Tr}^1$

The Deduction Theorem was formulated as S5, and uses not only the concept of consequence but also *saying that*.

Inf1. $\bigwedge Xqpa^r X \subset \mathrm{Tr} \wedge \sim(S\langle q\rangle \subset Cn(X)) \wedge S\langle p \supset q\rangle(a) \wedge Cn(S\langle p\rangle \cup X) - Cn(X) = Cn(S\langle q\rangle \cup X) - Cn(X) \wedge p . \supset q^1$

Proof. $\bigwedge Xqpa$, if

1.	$X \subset \mathrm{Tr}$	
2.	$\sim(S\langle q\rangle \subset Cn(X))$	
3.	$S\langle p \supset q\rangle(a)$	
4.	$Cn(S\langle q\rangle \cup X) - Cn(X) = Cn(S\langle q\rangle \cup X) - Cn(X)$	
5.	p, then	
6.	$S\langle q\rangle \subset Cn(S\langle q\rangle \cup X) - Cn(X)$	(Cn1, Cn2, 2)
7.	$S\langle q\rangle \subset Cn(S\langle q\rangle \cup X) - Cn(X)$	(6, 4)
8.	$S\langle q\rangle \subset Cn(S\langle q\rangle \cup X)$	(7)
9.	$S\langle p \supset q\rangle \subset Cn(X)$	(S5, 8)
10.	$Cn(X) \subset \mathrm{Tr}$	(Cn4, 1)
11.	$S\langle p \supset q\rangle \subset \mathrm{Tr}$	(9, 10)
12.	$\bigwedge a^r S\langle p \supset q\rangle(a) \supset : \mathrm{Tr}(a) \supset . p \supset q^1$	(Tr1)
13.	$S\langle p \supset q\rangle(a) \supset : \mathrm{Tr}(a) \supset . p \supset q$	(12)
14.	$\mathrm{Tr}(a)$	(11, 3)
15.	$p \supset q$	(13, 3, 14)
	q	(15, 5)

Inf2. $\bigwedge Xqpa^r X \subset \mathrm{Tr} \wedge S\langle q\rangle \subset Cn(X) \wedge S\langle p \supset q\rangle(a) \wedge p . \supset q^1$

Proof. $\bigwedge Xqpa$, if

1. $X \subset \mathrm{Tr}$

2. $S\langle q \rangle \subset \mathrm{Cn}(X)$

3. $S\langle p \supset q \rangle(a)$, then

4. $\mathrm{Cn}(X) \subset \mathrm{Tr}$ (Cn4, 1)

5. $S\langle q \rangle \subset \mathrm{Cn}(X) \cup \mathrm{Cn}(S\langle p \rangle)$ (2)

6. $S\langle q \rangle \subset \mathrm{Cn}(S\langle p \rangle \cup X)$ (Cn2, 5)

7. $S\langle p \supset q \rangle \subset \mathrm{Cn}(X)$ (S5, 6)

8. $S\langle p \supset q \rangle \subset \mathrm{Tr}$ (7, 4)

The rest of the proof proceeds as in the proof of Inf1.

Inf3. $\wedge Xpqa^\ulcorner X \subset \mathrm{Tr} \wedge S\langle p \supset q \rangle(a) \wedge \mathrm{Cn}(S\langle p \rangle \cup X) - \mathrm{Cn}(X)$
$= \mathrm{Cn}(S\langle q \rangle \cup X) - \mathrm{Cn}(X) \wedge p . \supset q^\urcorner$ (Inf1, Inf2)

Inf3 says that, in Interpretations 2, 0A2 holds under two additional conditions. The first requires that the assumptions be true; this amounts to selecting only some of the interpretations from the set of Interpretations 2. The second condition demands that there be a sentence which says that $p \supset q$. If $S\langle p \supset q \rangle(a)$, then there are b and c such that $S\langle p \rangle(b)$ and $S\langle q \rangle(c)$. But if $S\langle p \rangle(b)$ and $S\langle q \rangle(c)$, then there does not always have to be an a such that $S\langle p \supset q \rangle(a)$. In particular, if b and c belong to different sublanguages, the formation of the conditional may be ungrammatical or marginal. 'If X is an even number, then there is a liquid that dissolves X' is unacceptable or, at best, marginal.

In this interpretation, relative to a fixed X, 0A6 is

$$\wedge Y^\ulcorner \sim Y \alpha R(Y)^\urcorner$$

There is no set Y which is the information of a sentence which says that Y is Russellian. In other words, if Z is the information of a sentence which says that Y is Russellian, then $Z \neq Y$. (Here, if a set is Russellian and is the information of a sentence which says that p, then $\sim p$.)

NOTES

[1] Notational conventions. The upper corners show the scope of quantifiers. Parentheses embrace the arguments of a functor which stands before the left parenthesis. But

sometimes a familiar functor appears between the arguments; for instance the sign of implication. '$\supset$' will be placed between its arguments. Instead of a pair of parentheses around a sentential argument a dot is used in the way of Quine in his *Mathematical Logic*. Quantifiers $\wedge x^r \wedge y^r \ldots$'[11] are combined into $\wedge^r xy \ldots$'. The order of variables in the quantifier is that of their free occurrences in the scope of the quantifier. Proofs are written in Leśniewski's fashion, similar in some respects to proofs in natural deduction. Indentation in the proofs indicates that the statement is under the governing quantifier. For instance, in the proof of A8, 1 is under the quantifier '$\wedge af$', 2 and 3 are under '$\vee g$' as well, and the concluding line '$f(a)$' is again under '$\wedge af$' only.

[2] Gottlob Frege, *Grundgesetze der Arithmetik*, vol. ii, pp. 253–65; in English, *Translations from the Philosophical Writings of Gottlob Frege*, Peter Geach and Max Black (eds.), Oxford, 1952, pp. 234–44.

[3] The word 'argument' is ambiguous. Either it refers to an entity which is acted upon by a function, or it is a phrase which together with another phrase, namely, its functor, and perhaps with other arguments, forms a longer phrase. The distinction between a function and a functor (proposed by Kotarbiński and generally accepted in the Warsaw school of logic) does not have its parallel for 'argument'.

[4] See Alonzo Church, 1976, 'Comparison of Russell's Resolution of the Semantical Antinomies with That of Tarski', *The Journal of Symbolic Logic* **41**.

[5] Bolesław Sobociński. 'L'analyse de l'antinomie Russellienne par Leśniewski', *Methodos*, vol. I (1949), pp. 94–107; pp. 220–28; pp. 308–16; vol. II (1950), pp. 237–57. Sections 2, 3, and 4 of the present paper constitute a restatement of what is in Sobociński's paper. (Errata to Sobociński's paper: p. 226, line 2 from the bottom, put a left-hand parenthesis before 'D'; p. 238, line 18, instead of '=' put '≡'; p. 238, line 24, instead of 'et' put 'est'.)

[6] W. V. Quine, 1955, 'On Frege's Way Out', *Mind* **64**, pp. 145–59. Reprinted in Quine's *Selected Logical Papers*, Random House, New York, 1966, and in *Essays on Frege*, ed. by E. D. Klemke, University of Illinois Press, 1968.

[7] The formulations are mine. I owe much to conversations over a period of many years with Zellig Harris, Richard Smaby, James Munz, and, above all, the late Beverly Robbins. The criticisms by Richard Martin of my previous papers were very instructive. The last criticism, 'On Hiż's Sentential Holism', is in his *Logico-Linguistic Papers*, Foris, Dordrecht, 1981.

[8] In some of my previous papers ('The Role of Paraphrase in Grammar', *Monograph Series in Languages and Linguistics*, **17**, 1964; 'Aletheic Semantic Theory', *The Philosophical Forum* **1**, 1969) I drew the distinction between a weak and a strong semantics. A weak semantics does not assign semantic values (denotation, information, truth-value, etc.) to any phrase but asserts that two phrases have the same or a different semantic value. The present investigation belongs to a strong semantics. This does not mean that a weak, or weaker, semantics does not have a role to play. They are different enterprises, each legitimate.

[9] There are two main differences between Leśniewski's original proofs relevant to the issue and their rendering here. First, Leśniewski considered 'Kl' – for 'the class of' – as a functor forming a noun out of a noun. For him, '$A \in \mathrm{Kl}(b)$' says that A is the class of bs; and 'A', 'b' and '$\mathrm{Kl}(b)$' were all the same grammatical (syntactical and semantical) category. In the present paper, 'aaf' and '$a\langle f\rangle(a)$' are used in such a way that 'a' and 'f' are

of different grammatical categories, so that '$f(a)$' is of the category of a sentence. This necessitates a change in treating such Leśniewskian expressions as '$A \in \text{Kl}(B)$' where 'B' is thought of as a name of an individual. I changed it to '$a\,\alpha\,\iota b$' or to '$\alpha\langle\iota b\rangle(a)$', and similarly in other cases. This departure from Leśniewski's original way of stating his ideas is the result of framing the considerations in (2nd order or higher) functional logic rather than in Leśniewski's ontology. The second change, connected with the first, is in the use of definitions. Leśniewski accepted ontological definitions which are of the form

$$\wedge ab \ldots c \ldots \ulcorner a \in \dagger\langle b \ldots\rangle(c \ldots) \equiv . a \in a \wedge \varphi \urcorner$$

where '$\dagger$' is a sign new in its grammatical category and φ has just the $a, b, \ldots, c, \ldots$ as its free variables. The only definitions I use are the usual, "sentential" definitions where the definiendum is of the grammatical category of a sentence and is of the form

$$\dagger\langle a \ldots\rangle(b \ldots)$$

On the eliminability of ontological definitions see my "Ontological Definitions in Augmented Protothetics", an abstract, *The Journal of Symbolic Logic* **31** (1966), pp. 149–50. The arguments inside '$\langle\ldots\rangle$' are considered parameters; the groupings of parentheses is to the left. In '$\dagger\langle a\rangle(b)$', '$\dagger\langle a\rangle$' is the functor and '$b$' its argument: '$\dagger$' in turn is the functor in '$\dagger\langle a\rangle$' and 'a' its argument.

[10] Quine's paper cited in note 6. I do not attempt to compare Quine's and Leśniewski's proofs.

[11] To see Leśniewski's system in a broader context, consult the special issue of *Studia Logica* **36**, 4 (1977) devoted entirely to Leśniewski.

[12] Zelig Harris, *Mathematical Structures of Language*, Interscience Publishers, New York, 1968, pp. 145–46.

[13] Alfred Tarski, *Logic, Semantics, Metamathematics*, translated by J. H. Woodger, Oxford, 1956, p. 417.

[14] See p. 32 of Tarski's book quoted in Note 13, Axiom 8*. S5 here and Tarski's Axiom 8* do not say exactly the same thing. In S5 there is no syntactic device for referring to a sentence.

[15] About sublanguages consult the book, *Sublanguage: Studies of Language in Restricted Semantic Domains*, edited by Richard Kittredge and John Lehrberger, Walter de Gruyter, Berlin, 1982.

Dept. of Linguistics
University of Pennsylvania
Philadelphia, PA 19104
U.S.A.

RICHARD JEFFREY

DE FINETTI'S PROBABILISM

> One must invert the roles of inductive reasoning and
> probability theory: it is the latter that has autonomous
> validity, whereas induction is the derived notion. . . . I
> think that the way in which I have sought to resolve the
> problem of inductive reasoning is nothing but a trans-
> lation into logico-mathematical terms of Hume's
> ideas . . . On this view, every evaluation of probability
> has only a subjective value . . .
>
> Bruno de Finetti[1]

If de Finetti got it from Hume, then Hume got it from Carneades (ca.
200 B.C.), via Cicero, Sextus Empiricus, Montaigne, and Bayle.[2] It is
something of a pun to call it "probabilism" over the whole 2000 years
and more of that line, but you can see why it is so called by dipping into
an English translation of Cicero's Latin rendering (in a dialogue) of the
Greek of Carneades's student, Clitomachus:[3]

Carneades holds . . . that there is no presentation of such a sort as to result in knowledge
through perception, but many that result in a judgement of probability. For it is contrary
to nature for nothing to be probable, and entails that entire subversion of life of which
you, Lucullus, were speaking; accordingly even many sense-percepts must be deemed
probable, if only it be held in mind that no sense-presentation has such a character as a
false presentation could not also have without differing from it at all.

Here, irresistibly, "probable" translates Cicero's "*probabile*", i.e., a
technical philosophical term he introduced in *Academica* to translate
the Greek "*pithanon*", which means "persuasive", in nontechnical
English. This "*probabile*" is a prequantitative forerunner of the notion
of subjective probability that de Finetti deployed, and associated with
Hume.[4]

Here is a traditional cartoon of the Hellenistic background.[5] The
Stoics sought the criterion whereby true presentations (*phantasia*) may
be recognized, and some (e.g., Chrysippus) thought to have found it in
the apprehensive (*cataleptic*) character of (some) presentations that
originate in reality: such a presentation grabs you, and compels assent.

Synthese **60** (1984) 73–90.   0039–7857/84/0601–0073 $01.80
© 1984 *by D. Reidel Publishing Company*

(Or perhaps you grab *it*. The Greek is ambiguous.) Carneades denied the existence of any such criterion. In particular, he rejected the candidates for the role that the Stoics had canvassed, e.g., such intrinsic qualities as clarity and distinctness. He held (and this was his "probabilism") that we can, and must, make do with uncertified probabilities, i.e., qualified judgements of which cataleptic assent is a polar case.

The Stoic ideal was the sage, or wise man. Cicero's dialogue continues:

Thus the wise man will make use of whatever specimen of[6] probable presentation he encounters, if nothing presents itself that is contrary to that probability, and his whole life plan will be charted out in this manner. In fact even the person whom your school brings on the stage as the wise man follows many things probable, that he has not grasped nor perceived nor assented to but that possess verisimilitude; and if he were not to approve them, all life would be done away with.

And a wise man may well make a sea voyage, even in the absence of a cataleptic presentation of safety, but merely because the weather is good, the distance short, *etc*.[7]

He will therefore be guided by presentations of this sort to adopt plans of action and of inaction.... For he is not a statue carved out of stone or hewn out of timber; he has a body and a mind, a mobile intellect and mobile senses, so that many things seem to him to be true, although nevertheless they do not seem to him to possess that distinct and peculiar mark....

Now read a bit of Hume. He begins Part IV of Book I of his *Treatise of Human Nature* with a consideration 'Of skepticism with regard to reason'. He holds that "all knowledge resolves itself into probability", and then considers the "turtles all the way down" effect:[8]

When I reflect on the natural fallibility of my judgement, I have less confidence in my opinions, than when I only consider the objects concerning which I reason; and when I proceed still farther, to turn the scrutiny against every successive estimation I make of my faculties, all the rules of logic required a continual diminution, and at last a total extinction of belief and evidence.

He finds this argument unanswerable but, still, ignorable: concerning such total skepticism he says that

neither I, nor any other person was ever sincerely and constantly of that opinion. Nature, by an absolute and uncontroulable necessity has determined us to judge as to breathe and feel; nor can we any more forbear viewing certain objects in a stronger and fuller light, upon account of their customary connection with a present impression, than we can hinder ourselves from thinking as long as we are awake, or seeing the surrounding bodies, when we turn our eyes toward them in broad sunshine.

He concludes (in italics)

that all our reasonings concerning causes and effects are derived from nothing but custom; and that belief is more properly an act of the sensitive, than of the cogitative part of our natures.

It is the second part of this conclusion that I identify with de Finetti's probabilism,[9] and with Carneades's, as reported by Cicero, above, and by Sextus Empiricus.[10]

I take de Finetti's probabilism to be in part a response to the sort of difficulty with skepticism that Montaigne reports:[11]

I can see why the Pyrrhonian philosophers cannot express their general conception in any manner of speaking; for they would need a new language. Ours is wholly formed of affirmative propositions, which to them are utterly repugnant; so that when they say "I doubt," immediately you have them by the throat to make them admit that at least they know and are sure of this fact, that they doubt.

The new language is nothing else than the probability calculus; but de Finetti's twist is to reject such questions as the following: "You say that the probability of A is x. But what is the probability of *that*?" He says:

Any assertion concerning probabilities of events is merely the expression of someone's opinion and not itself an event. There is no meaning, therefore, in asking whether such an assertion is true or false or more or less probable.[12]

But as Brian Skyrms points out,[13] if it is possible for us to discuss our beliefs (as de Finetti certainly supposes), there must be propositions of the form, "R.C.J.'s degree of belief in A is x", to which you can assign probabilities, and to which I, too, can assign probabilities between 0 and 1 if I am less than clear about what I think.

Still, I take de Finetti's position to be quite intelligible: his point is that probabilities are states of mind, e.g., for me to attribute probability $\frac{1}{2}$ to rain tomorrow is for me to think either side of an even-money bet on rain tomorrow equally attractive, and to have various other such practical attitudes. But he sees no practical attitudes that would similarly display my degree of belief in my degree of belief in rain tomorrow's being $\frac{1}{2}$; nor does he think that a deficiency, for one's concern as an active agent is to choose coherently, i.e., in accordance with one's actual graded beliefs and desires. He has been at pains to provide methods whereby these can be accurately elicited,[14] and I suppose that in the context of such elicitation he would have to admit that (1) uncertainty about the exact numerical values is real and even quantifiable, in principle – but that (2) the effort it would take to quantify those second-order uncertainties would be better expended in closer examination of the values of the first-order ones, which connect directly to practical life.

The fact is, of course, that one *can* have turtles all the way down, without thereby losing probabilities altogether.[15] But the fact is, too, that the principal objects of interest are the ground-level probabilities, unknown or unformed though they be. I shall return to these questions shortly. But first, there is another, historiographical obstacle to be cleared away, i.e., the argument in Ian Hacking's [1975] book, *The Emergence of Probability*, that our very concept of probability is no older than the 17th century. If so, the associations I have been making with Carneades and the middle academy would be quite specious. Here, I lean heavily on the work of Daniel Garber and Sandy Zabell [1979].

This is Hacking's thesis:[16]

Probability has two aspects. It is connected with the degree of belief warranted by evidence, and it is connected with the tendency, displayed by some chance devices, to produce stable relative frequencies. Neither of these aspects was self-consciously and deliberately apprehended by any substantial body of thinkers before the time of Pascal.

This is the thesis of the emergence of our very *concept* of probability only in the 17th century – a thesis that Hacking floats in large part in order to explain the explosion of mathematical probability theory in the second half of that century. As is well known, the first monograph on the subject was Huygens's 'On calculating in games of luck' (1656–1657), which remained the basic textbook well into the next century. There has been essentially nothing in the way of a mathematical theory of probability before ca. 1654 (i.e., the beginning of the cor-respondence between Fermat and Pascal that Huygens was reporting upon, in his monograph), but in 1713, with the publication of James Bernoulli's *Art of Conjecturing*, our own era of laws of large numbers had begun. And yet, as Hacking points out,[17] the mathematical tools used by Huygens and his immediate successors were of a very elemen-tary character, so that he wonders why the mathematical theory did not begin earlier, if the prequantitative probability concept had been in existence since antiquity.

But it strikes me as gratuitous to postulate special creation of the concept of probability in particular, at a time when quantification was coming to be the name of the scientific game quite generally. Galileo had shown the power of the method in physics, Descartes had done it in geometry, and in 1665, according to Newton's recollection of the period,[18]

in November [I] had the direct method of fluxions [differential calculus], and in the next year in January had the theory of colours, and in May following I had entrance into ye inverse method of fluxions [integral calculus]. And the same year I began to think of gravity extending to ye orb of the moon, and . . . from Kepler's Rule of the periodical times of the Planets . . . I deduced that the forces wch keep the Planets in their orbs must [be] reciprocally as the squares of their distances from the centers about which they revolve: and thereby compared the force requisite to keep the Moon in her Orb with the force of gravity at the surface of the earth, and found them answer pretty nearly. All this was in the two plague years of 1665 and 1666, for in those days I was in the prime of my age for invention, and minded Mathematics and Philosophy more than at any time since.

And the *dramatis personae* in the beginning of mathematical probability theory were closely involved with the mathematicization of physics: Fermat, Pascal, and, most notably, Huygens (the priority of whose work on centrifugal force was acknowledged by Newton).[19] The half century in question was a time of intoxication with the power that mathematics gave the human mind. The wonder would have been for probability *not* to have been mathematicized then.

Then I rejected Hacking's reasons for supposing that "the modern" concept of probability did not exist, even unmathematicized, before the 17th century. But I also reject his identification of "that" concept with a certain Janus-faced frequentist-inferential pushmi-pulyu.[20] The nature of probability has been a vexed question throughout the three centuries and more of development of the mathematical theory. It is only in our century, with the advent of abstract measure theory, that it has been possible to separate clearly the mathematical theory, on which agreement in general,[21] from the philosophical dispute[22] about the nature of probability. To identify "our" concept of probability with Hacking's hybrid is to seek to end the dispute by declaring the two most popular contenders to be joint winners.

I prefer de Finetti's probabilism, as giving a coherent account of the two "aspects" that Hacking points to, while making sense of the historical facts. Certainly, Huygens's (1656) viewpoint is close to that of de Finetti, for whom the probability of a proposition is the (buying-or-selling) price of a chance to receive 0 or 1 depending on whether the proposition is false or true.[23]

Huygens begins[24] by asking what it is worth, to have equal chances of receiving two amounts. (*We* would say that the question concerns the value of a bet on a proposition of probability $\frac{1}{2}$.) Calling the unknown value "x", he observes that with x in hand instead of the bet, he is in a position to bet afresh, against someone who also has x, so that the winner will have $2x$ and the loser will have nothing. To discover the value of x, he observes that the worth of the bet would be no different if it were agreed that the winner would pay the loser a small consolation prize a, so that after all, the loser has a and the winner has $2x - a$. Thus, to have equal chances of having a or $2x - a$ is worth x. Or, as Huygens puts it (changing notation by setting $b = 2x - a$):

PROPOSITION I. To have equal chances of having a or b is worth $\frac{a + b}{2}$ to me.

Hacking is well aware of all this: his account of Huygens's reasoning[25] is lucid and sensitive; and he says this, concerning Huygens's treatment of the problem solved in proposition I (and in other generalizations of that problem):

His solution to this has a singularly modern flavour because it has been revived, in a different format, in the personalist theory of F. P. Ramsey and L. J. Savage.

And later:

The fair prices worked out by Huygens are just what we would call the expectations of the corresponding gambles. His approach made expectation a more basic concept than probability, and this remained so for about a century. There is nothing wrong with this practice: it has been elegantly revived by P. Whittle [1970].

But in these passages, Hacking presents as merely quaint and fortuitous what I see as a telling affinity between 20th century subjectivism and the shape in which probability theory emerged in the 17th century. Along with P. Whittle, Hacking could have mentioned Bruno de Finetti,[26] as one who sees probability as simply a special case of mathematical expectation. And before even Ramsey and Savage, Hacking could have pointed to de Finetti as one whose format is practically the same as Huygens's. This last excerpt from Hacking's book continues:

Huygens does not use any long-run justification for his "fair prices". Perhaps this is partly because averages were not well established as a natural way of representing data. . . .

But perhaps this is because Huygens, like de Finetti, saw the single-case justification as precisely what was wanted.

It seems to me that Hacking is reading the early history of mathematical probability theory with an anachronistic bias toward the frequentist-inferential pushmi-pulyu. As an antidote, I have been urging a reading with a different bias, which I think less anachronistic (e.g., because it yelds what I find to be a intelligible reading of Carneades and Huygens), and because it seems to me that de Finetti has shown how that sort of probabilism puts frequencies and inferences intelligibly into the picture.

For as de Finetti points out, early[27] and late,[28] an intimate connection between probability and expectation appears at the most elementary and general level, as an immediate consequence of the basic laws of probability, i.e., those valid for all probability measures whatever – even those for which no laws of large numbers are provable. To see this, let $A_1, \ldots, A_n$ be a finite sequence of propositions *of any sort*. At one extreme, the A_i might all be identical, so that we have n repetitions of some such proposition as *the ace will turn up on the first roll of this die*. At the other extreme, the A_i might have widely different subject-matters and probabilities, e.g., A_1 might be the foregoing proposition about the die, A_2 might be the proposition that there will be some rain in San Francisco during September of the year 1989, *etc*. Let P be a probability function satisfying the basic laws of probability, and let E be the corresponding expectation function. We know that if f_i is the indicator function of the proposition A_i, *i.e.*, if $f_i(w)$ is 0 or 1 depending on whether A_i is false or true in "world" w, then probability is the expectation of the indicator:

$$P(A_i) = E(f_i)$$

Then the probability of a proposition is always the dollar value (subjectively fair buying-or-selling price) of a one-dollar bet on that proposition. ($f_i(w)$ is the payoff from that sort of bet on A_i in world w.) Summing over the n values of i, and then using the fact that expectation is additive, we have

$$P(A_1) + \ldots + P(A_n) = E(f_1) + \ldots + E(f_n) = E(f_1 + \ldots + f_n)$$

As $f_1(w) + \ldots + f_n(w)$ is the number of truths among the A's in world w, we can summarize in English:

Sum of probabilities = Expected number of truths

Or, dividing each side of the equation by n, and then applying the linearity of E in the form $(Ef)/n = E(f/n)$, we have the law

(*)   *Average probability = Expected truth frequency*

i.e., in symbols,

$$\frac{P(A_1) + \ldots + P(A_n)}{n} = E\frac{f_1 + \ldots + f_n}{n}$$

Now let us see what this basic relationship between probabilities and *expectation* of frequency implies about probabilities and experience of frequency.

In case P is someone's personal or subjective probability, i.e., someone's judgmental probability as it exists at a certain time, we can consider the case where the person has become convinced that the (relative) frequency of truths among the A's is r, but has no particular information about *which nr* of the A's are true. At that time, i.e., relative to the probability function P that characterizes his state of belief then, his expectation of truth frequency (the right-hand side of equation (*)) will be precisely r, i.e., the value assumed by $(f_1(w) + \ldots + f_n(w))/n$ when w is the real world. Thus, concerning subjective probabilities, (*) says:

> *If one knows the actual truth frequency, then average probability = actual truth frequency.*

The most striking case is that in which the subject is convinced that the A's are equiprobable, i.e., the case in which any probability measure P that adequately represents his frame of mind must assign equal probabilities to the A's:

$$P(A_1) = \ldots = P(A_n)$$

In that case we must have

$$P(A_i) = r$$

for all i from 1 to n. Thus, as de Finetti points out, relative frequencies do determine probabilities in a basic species of cases, when probabilities are understood subjectivistically. The species is that in which one knows the relative frequency, and takes all probabilities to be equal. In terms of A's that report "success" on different trials of an experiment, what is assumed is that we know the relative frequency of success, and are convinced merely that the probability of success is constant from trial to trial. What we do *not* assume it that the trials are independent, or exchangeable, *etc*.[29] Nor do we assume that the number of trials is very large. Nor is the equality between $P(A_i)$ and r asserted only approximately. In (*) we have no law of large numbers: it is a law of *all* finite (i.e., "small") numbers. Therefore (with an exuberance that de Finetti would deplore[30]) I dub it

DE FINETTI'S LAW OF SMALL NUMBERS: Always, average probability = expected truth frequency. Then in the case where the overall success rate r is known, in an experiment where the subjective probability p of success is constant from trial to trial, $p = r$, i.e., probability = success rate.

The law (*) of small numbers covers the case where you know the relative frequency of "success" in the whole population of "trials". The case, where you learn the relative frequency for the first n trials and then consider the next trial, or the next m, require more assumptions, e.g., exchangeability. I shall not spend time here on de Finetti's account of that aspect of the relationship between probability and frequency, for it has come to be well known even outside the ranks of subjectivists via de Finetti's representation theorem,[31] according to which any symmetric probability measure P is uniquely expressible as a weighted average of binomial probability measures B_p, where the weighting function can be thought of as identifying the probabilities that the unknown constant value p of success on every trial lies in this or that part of the unit interval. (De Finetti proves the theorem in more general forms, e.g., for multinomial distributions, and for random variables.[32]) The representation theorem is of interest as providing a subjectivistic interpretation of objectivistic talk about unknown objective chances, as well as for its implications for "prevision" of frequencies, according to which one's probability measure after learning the relative frequency of success in the first n trials is symmetric if the initial one was, with the weighting function for the new symmetric measure pushing more of the "mass" of the distribution toward the observed relative frequency than was there originally – provided the original distribution function for p had no "flat spots" of positive length.[33]

My main demurral from (or gloss upon) de Finetti's line has to do with the contrast between elicitation of probability judgments when thought of as a matter of introspecting a preexisting state or disposition, and as a matter of forming such a state or disposition on the spot. I incline toward the latter view. To see why (in a quick, caricaturish way), consider the simplest version of the preexistence model, in which one's cognitive state (of graded belief) at any time is describable by a probability measure on all the propositions that are expressible in one's language. That would make grotesque demands upon memory capacity. Thus, as Gilbert Harman is fond of pointing out, with as few as 300 assorted propositions, the information required to determine the probabilities of all $2^{2^{300}}$ of their truth functional compounds needs to be stored in the form of $2^{300} - 1$ values, in the worst case. But that's a lot larger than the number of molecules that physicists say there are in the universe. There is a first pass at the "combinatorial explosion" argument against a version of subjectivism that has us storing full-blown probability measures in our heads, and updating them as new data flow in through the senses.

The argument is far from decisive, e.g., it might be taken to show that although the thing is perfectly possible, it requires a Carnapian approach, in which the probability function is analytically formulable, so that the $2^{2^{300}}$ values are computable on demand (i.e., those that are needed) from a memory of the conjunction e that represents all of one's evidence to date.[35] Updating is then a matter of adding more conjuncts to E.

Then the combinatorial-explosion argument need not bear against the stored-probability-function picture. But of course it *is* suggestive, and may serve to reenforce a tendency one has anyway, upon noticing that we do not really remember a lot of middling probability values, to think of the probability function as something we work up as occasions demand, so that we remember (at most) fragments of it, or parameters or features that determine chunks of it. Item: where we take a process to be binomial, with constant probability p of success and with trials independent, the effect of storing the whole $2^{2^{300}}$ probabilities regarding 300 trials can be had just by remembering the number p and the fact that the process is binomial. We do that sort of thing with coins and dice and sexes of babies, even cheating a bit because we prize the efficiency of that sort of encoding enough to accept a bit of distortion in the picture of our real mental state. Notice that two-valued probability functions are guaranteed to be binomial, and that is why truth values (or zeros and ones) are so welcome, as items to store – and why we are prepared to cheat a bit there, too, in the interest of efficiency. Remember: the combinatorial explosion doesn't have to be as bad as the one Harman mentions to be bothersome. It is true that the specialized systems in our heads (vision, speech, etc.) can handle lots of information of special sorts very fast; but for general doxology, where we want to allow for inventiveness and response to situations that evolution hasn't slanted us toward via actual circuitry, we need to keep information accessible and quickly surveyable.

So Harman's objection tends to promote the move away from introspection of preexisting probability functions, toward formation of probability functions ad hoc. But let me reiterate that the combinatorial-explosion argument is nowhere near *forcing* such a move, e.g., because it is framed in terms of utter precision, whereas the bothersome (un-Carnapian) computer model demands a limited number of decimal places of accuracy, as a corollary of the fact that the number of bits of storage is limited. *Ability to store a single arbitrary real number in the unit interval already requires no end of hardware* (or wetware), albeit in a single register. Bearing that in mind, we see that there is no combinatorial explosion if utter precision is postulated, for the same denumberable infinity of bits of storage that would accommodate 300 probability values will equally well accommodate $2^{300} - 1$, or (equally well) the whole $2^{2^{300}}$ values.

But quite apart from combinatorial explosions, there is plenty of reason to think in terms of probability functions that are rigged up as occasions demand them. For example, if one expects to learn what the success rate has been before one will need to say what the probabilities are of successes on the several trials, then one does well to put off the business of eliciting probabilities until the empirical value of r is known. The law of small numbers then imposes the constraint

$$P(A_1) + \ldots + P(A_n) = rn$$

on the n probabilities of success, in the light of which one might find that further data (of an inward- or outward-looking sort) help fix the values of the $P(A_i)$ with some precision. (In the extreme case, a judgment of equiprobability fixes $P(A_i) = r$ for all i.)

Of course, the law of small numbers will take us only so far. For stronger results along those lines we need to make special assumptions about P, e.g., assumptions that make the variance small, so that the owner of the belief function P will regard himself as well-calibrated.[36] But the law (*) itself remains of interest as an antidote to the impression, easily gained in the study of mathematical probability theory, that it is only in special circumstances – notably, those in which laws of large numbers hold – that probability theory can be brought to bear on the frequentist-inferential complex that Hacking identifies as our modern concept of probability.

With de Finetti, I take the probability calculus to have an autonomous validity: far from resting on frequentist or inferential foundations, it serves as a framework intelligibly connecting frequentist and inferential aspects of our thinking. The credentials of that framework come in part from the consideration that where its basic constraints are violated, behavior under uncertainty can be gratuitously self-defeating, i.e., self-defeating in ways in which it cannot be if those constraints are respected.[37] De Finetti quotes Poincaré[38] on the ineluctability of probabilistic reasoning: the probability calculus (Poincaré said)

rests on an obscure instinct, which we cannot do without; without it science would be impossible, without it we could neither discover a law nor apply it.

Perhaps so. But remember that probability theory is not descriptive psychology on the model of psychophysics, e.g., no one thinks there are localized potentials in the brain that constitute the physiological underpinnings of the subjective probabilities that manifest themselves in our behavior. In the computer analogy, it is the program, not the circuitry, that "houses" such probabilities as we entertain. If there is an underlying instinct, there is also a large component of cultural artifact. Probability attributions are far from being state parameters of the human organism, measurable by oral credometers.

Whatever else they may be, probability ascriptions *are* recognizable features of our preference rankings, i.e., of our detailed decision-making policies – to the extent that we have them and they hang together.[39] Thus, they feature in the determination of buying-or-selling prices for chancy arrangements. In that sense, the subjectivistic theory is canonical, e.g., even if you insist that you are talking about objective chance when you say the probability of throwing an ace next is 1/6, that probability ascription will *also* cash out as a $1 buying-or-selling price for a ticket worth $6 or nothing depending on whether or not the ace turns up next.[40] As an account of the pragmatics of probability, subjectivism sweeps the field. And the syntax (say, the standard measure-theoretical setup) is fairly unproblematical. The radical part of de Finetti's claim is that *there is no "objective" semantics for probability*: probability statements have no truth values when viewed as statements about the things to which probabilities are attributed.

But this claim does not have all the alarming implications that first come to mind. The claim is, e.g., that if you assert that the probability of drawing a face card first is 1/2, and I say it is 12/52, we are not thereby disagreeing about something in the world outside us, as we would be if you said that the deck in question is a pinochle deck (i.e., 48 cards, 24 of which are face cards), and I said that it is a normal 52-card deck (i.e., with 12 face cards). But to claim that is not to deny that there are facts about the outside world (say, the kind of deck, and how – and if – it has been shuffled), agreement on which would induce agreement about probabilities. The claim is simply that our probability assertions are not conflicting assertions about such facts, but only assertions about our current decision-making policies – policies whose undoubted sensitivity to information about such facts is not packed into the meaning or the truth conditions of the relevant probability assertions.

Nor does the claim have the implications that Frank Ramsey[41] attributed to it, i.e., that our differing probability attributions reflect no more real disagreement than do our differing statements when one of us says, "I went to Grantchester yesterday," and the other says, "I did not." Differences between probability attributions by different people are likely to represent practical disagreement when it comes to cooperative action, whereas Ramsey's Grantchester example is purely theoretical by design. (But even there, there might be real disagreement, in case we agree that we had spent all of yesterday together.) Perhaps the right thing to say here is that the bald, differing probability statements do not conflict in any intrinsic way, but that the conditions of human life (especially, the need for cooperation in many important circumstances) are likely to produce practical conflict in such cases, even though there is no strict theoretical conflict.

The radical claim that de Finetti makes, and characterizes as a translation of Hume's ideas into logico-mathematical terms, is that this subjectivistic concept of probability is all we need for science, for statistics, and for decision making under uncertainty. He is most anxious to deny that at any of these points we need an objective concept of probability: real propensities, objective chances, or whatever. But again, the claim is less alarming than it sounds, for de Finetti is far from denying the importance of relative frequencies – as his law of small numbers testifies. Rather, his insistence is that relative frequencies are *not* probabilities, although our attitudes about them are connected quite tightly *and obviously* with our probability attributions. He insists on the superficiality of that connection: it is there for all to see who know that expectations are linear, and that probabilities are expected values of indicator functions. Of course, the law of small numbers is mathematically trivial. I call it a "law" simply to emphasize the fact that probability and relative frequency are two, not one: no consequence of the law of identity, "$p = r$" is a consequence of the law of small numbers in the presence of futher conditions that give the notation "p", "r" its sense.

De Finetti's radical claim is that a subjectivistic concept, semantically ungrounded in external reality, is the only concept of probability that we need for our theoretical and practical purposes. Early in this essay I tried to connect that claim with issues, and with claims (also called "Probabilism") that were debated in the middle Academy (e.g., ca. 150 B.C.), and which explosively reentered the European intellectual scene in the 16th century with the reemergence of Sextus Empiricus's "Outlines of Pyrrhonism". I suggested that *this* is the archeological layer most squarely underlying the mathematical probability theory that emerged in the second half of the 17th century; that the relevant Greek is Carneades, not Aristotle; and that the connection Hacking traces from *opinio*[42] to probability is an epiphenomenon of the real process, from people like Carneades through Cicero and Sextus to Montaigne and Pascal and Huygens, and to Bayle, and thence to Hume, and so on to us: to de Finetti.

NOTES

[1] De Finetti [1938]. This excerpt is from a translation in Jeffrey [1980], p. 194.
[2] See Popkin [1979], Chapter II, for an account of the 16th century revival of Greek skepticism, and Popkin [1955] for the Bayle-to-Hume connection. For sending me to

Carneades, I thank Sandy Zabell; for Bayle, Edwin McCann. My reading of Cicero and Sextus took place in a seminar of Michael Frede's, from which the present account of Carneades primarily derives. But see also Striker [1980].

[3] H. Rackham's translation of Cicero's *Academica*, II, 99. Carneades was a non-publisher.

[4] In medieval philosophical Latin, "*probabile*" commonly did duty for Aristotle's "*endoxon*", i.e., "generally believed", as in the passage from William of Ockham on pages 44–45 of Garber and Zabell [1979]. (As Michael Frede points out, the usage goes back at least as far as Boethius; and as Miles Burnyeat points out, in the early rhetorical work *De Inventione*, I, 46, Cicero uses "*probabile*" for "That . . . which for the most part usually comes to pass, or which is a part of the ordinary beliefs of mankind, or which contains in itself some resemblance to these qualities . . . ".) One should bear the *pithanon* alternative in mind when evaluating Hacking's ([1975], Chapter 3) claim that before the 17th century, *probabile* was primarily a qualification of opinion, and had the force of "widespread" (as in "*exdoxon*"), not of "*wahrscheinlich*". It was the reemergence of Sextus in the mid-16th century that provided the background in the light of which one now sees "*probabile*" as Latin for "*pithanon*" in *Academica*.

[5] For a likelier account, see Frede [1983].

[6] Rackham has "apparently" instead of "specimen of" for Cicero's "*specie*". The emendation is Michael Frede's.

[7] Montaigne translates this (pretty nearly) in his "Apology for Raymond Seybond" (1588): "Moreover, there is no sect that is not constrained to permit its sage to conform in a number of things that are not understood, or perceived, or accepted, if he wants to live. And when he goes to sea, he follows this course, not knowing if it will be useful to him, and relies on the vessel being good, the pilot experienced, the season suitable – merely probable circumstances. He is bound to follow them . . . He has a body, he has a soul; his senses impel him, his mind stirs him. Even though he does not discover in himself that peculiar and singular mark of the right to judge, and perceives that he must not pledge his consent, since there may be some falsehood resembling this truth, he does not fail to carry on the functions of his life fully and comfortably." (Translation by Donald M. Frame, p. 374.)

[8] "The earth rests on a giant elephant, which is standing on a giant turtle. After that, it's turtles all the way down." Hume's worry was echoed by Bertrand Russell, [1948], p. 416. For one rebuttal, see Hans Reichenbach, [1952], pp. 151–152.

[9] I espoused this sort of probabilism in Jeffrey [1968]: see sec. 3, "Belief: Reasons vs. Causes".

[10] See, e.g., *Against the Logicians*, I, lines 159–190.

[11] In the Donald M. Frame translation, p. 392.

[12] De Finetti [1972], p. 189.

[13] Skyrms, [1980a], Appendix II, and [1980b].

[14] De Finetti [1972], Chapters 3 and 4, and [1974], Chapter 5.

[15] See notes 7 and 12.

[16] Hacking, [1975], p. 1.

[17] Hacking, [1975], p. 6.

[18] From Gillispie, [1960], pp. 119–120.

[19] Gillispie, [1960], pp. 120–121. For accounts of the roles of Fermat, Pascal, and

Huygens in the emergence of physics, see Dugas [1958].

[20] Lofting [1922], Chapter 10.

[21] Or, anyway, orthogonal to the philosophical disagreement.

[22] So to characterize it is not to identify the disputants as professional philosophers: they are as likely to be mathematicians.

[23] Here the unit is the total stake, contributed by both bettors. This definition underlies the "Dutch book" argument in de Finetti [1937]: see the top of page 62 in Kyburg and Smokler [1980].

[24] 'Van Rekeningh in Spelen van Geluck' (1656–1657). The account here is from the French translation in Huygens [1920], p. 63. What I report here as Huygens's proof is characterized by him as a way of *discovering* proposition I. He continues: "The proof is easy. Indeed, possessing $\frac{a+b}{2}$, I can bet this sum against another player who also has $\frac{a+b}{2}$, and arrange with him that the winner gives a to the other. In this way I shall have an equal chance of having a if I lose, or b if I win; for in this latter case I obtain the stake $a + b$ and I give him a."

[25] Hacking, [1975], Chapter 11. The following excerpts are from pages 95 and 97.

[26] It is a commonplace that mathematically, probabilities are indistinguishable from expectations of indicator functions. De Finetti's enthusiasm for the expectation point of view reflects his philosophical stance (e.g., on p. xix of de Finetti [1972], where he pointedly uses "P" both for "*price*", i.e., expectation, and "probability".) Huygens's own discussion of his proposition I makes no reference to probabilities except as prices ("worths") of random variables.

[27] De Finetti [1937], Chapter II.

[28] De Finetti [1974], sec. 5.8.

[29] See de Finetti [1937] for exchangeability ("equivalence"). The version in Kyburg and Smokler [1980] is especially useful for its additional remarks by de Finetti.

[30] "In order that correct use be made of this theorem, we must make very clear that it is essentially trivial: otherwise, we run the risk of goodness knows what being read into it." (De Finetti [1974], p. 202.) The same can be said of Bayes's Theorem.

[31] *Symmetry* (Carnap's (1950) term) is the attribute of probability measures that is correlative with the attribute *exchangeability* of sets of propositions.

[32] De Finetti [1937], Chapter IV.

[33] See Gaifman [1971]: flat spots of positive length would violate condition (*) on page 245 there.

[34] E.g., see Harman [1980].

[35] Things are distinctly messier when my "probability kinematics" scheme is used, e.g., as in Jeffrey [1968], sec 2.

[36] Calibration is a matter of proximity between (1) the average of the probabilities attributed to a set of propositions, and (2) the *actual* relative frequency of truths among those propositions. Then the law of small numbers does not imply that owners of probability functions always regard themselves as well calibrated. If we use the square of the deviation between truth frequency and average probability as a measure of lack of calibration, then it is the expected value of that quantity, i.e., *the variance of the truth frequency*, that measures the degree to which the owner of the probability of expectation

function regards himself as uncalibrated. (Note that calibration, as defined here, is always relative to a set of propositions. And indeed, it seems unpromising to expect calibration to be a topic-neutral attribute of persons.) See Lichtenstein *et al.* [1977] for a survey of recent work on calibration.

[37] See the "Dutch book" argument in de Finetti [1937], Chapter I.

[38] At the beginning of the introduction to de Finetti [1937].

[39] For one account of probabilities as features of preference rankings, see Jeffrey [1965].

[40] If we think of (interpersonal) *market* prices, idiosyncratic nonlinearities of individual utilities (as functions of prices) are smoothed out.

[41] Ramsey [1931].

[42] Hacking [1975], Chapters 3, 4, 5.

REFERENCES

Burnyeat, Miles (ed.): 1983, *The Skeptical Tradition*, Berkeley and Los Angeles.

Carnap, Rudolf: 1962, *Logical Foundations of Probability* Univ. of Chicago Press, Chicago. Carnap, Rudolf and Richard Jeffrey (eds.), *Studies in Inductive Logic and Probability*, vol. 1, Univ. of California Press, Berkeley, 1971.

Cicero: 1933, *De Natura Deorum and Academica*, trans. by H. Rackham, Heinnem, London.

Dugas, Rene: 1958, *Mechanics in the 17th Century*, Central Book Co., Neuchatel.

de Finetti, Bruno: 1937, La prévision: ses lois logiques, ses sources subjectives', *Annales de l' Institut Henri Poincaré* **7**, 1–68. English translation in Kybury and Smokler [1980].

de Finetti, Bruno: 1938, 'Sur la condition d'équivalence partielle', *Actualites Scientifique et Industrielles* **739**, 5–18. English translation in Jeffrey [1980].

de Finetti, Bruno: 1970, *Teoria Delle Probabilità*, Torino, English translation, *Theory of Probability*, J. Wiley, New York, 1974 (vol. 1), 1975 (vol.2).

de Finetti, Bruno: 1972, *Probability, Induction and Statistics*, J. Wiley, New York.

Frede, Michael: 1983, 'Stoics and Sceptics on Clear and Distinct Impressions', in *The Skeptical Tradition*. See Burnyeat, 1983.

Gaifman, Haim: 1971, 'Applications of de Finetti's Theorem to Inductive Logic', *In Studies in Inductive Logic and Probability*. See Carnap and Jeffrey, 1971.

Garber, Daniel and Sandy Zabell: 1979, 'On the Emergence of Probability'. *Archive for History of Exact Sciences* **21**, 33–53.

Gillispie, Charles C: 1960, *The Edge of Objectivity*, Princeton Univ. Press, Princeton.

Hacking, Ian: 1975, *The Emergence of Probability*, Cambridge Univ. Press, Cambridge.

Harman, Gilbert: 1980, 'Reasoning and Explanatory Coherence', *American Philosophical Quarterly* **17**, 151–157.

Hume, David: 1739, *A Treatise of Human Nature*, Vol. 1, London.

Huygens, Christiaan: 1920, *Oevre Completes*, vol. 14 La Haye.

Jeffrey, Richard: 1965, *The Logic of Decision*, McGraw-Hill, New York. 2nd edition, revised, University of Chicago Press, 1983.

Jeffrey, Richard: 1968, 'Probable Knowledge', in Imre Lakatos (ed.), *The Problem of Inductive Logic*. North-Holland Pub. Co., Amsterdam. Reprinted in *Studies in Subjective Probability* See Kyburg and Smokler, 1980. (ed.) *Studies in Inductive Logic and Probability*, vol. 2, Univ. of California Press, Berkeley.

Kyburg, Henry Jr., and Howard Smokler (eds.): 1980, *Studies in Subjective Probability* (second edition, revised) J. Wiley, Huntington, NY.

Lichtenstein, Sarah, Baruch Fischhoff, and Lawrence Phillips: 1977, 'Calibration of Probabilities: The State of the Art', in Helmut Jungermann and Gerard de Zeeuw (eds.), *Decision Making and Change in Human Affairs*, D. Reidel, Dordrecht, pp. 275–324.

Lofting, Hugh: 1922, *The Story of Doctor Dolittle*, London.

de Montaigne, Michel: 1948, *The Complete Essays*, translated by Donald M. Frame, Stanford.

Popkin, Richard H: 1955-56, 'The Skeptical Precursors of David Hume', *Philosophy and Phenomenological Research* **16**, 61–71.

Popkin, Richard H: 1979, *The History of Skepticism from Erasmus to Spionza*, Univ. of California Press, Berkeley, 1979.

Reichenbach, Hans: 1952, 'Are Phenomenal Reports Absolutely Certain', *The Philosophical Review* **61**, 147–159.

Ramsey, Frank P.: 1931, *The Foundations of Mathematics and Other Logical Essays*. Routledge and Kegan Paul, London.

Russell, Bertrand: 1948, *Human Knowledge: Its Scope and Limits*, Simon and Schuster, New York.

Schofield, Malcolm, Myles Burnyeat, and Jonathan Barnes (eds.): 1980, *Doubt and Dogmatism: Studies in Hellenistic Epistemology*, Clarendon Press, Oxford.

Sextus Empiricus: 1933, *Outlines of Pyrrhonism*, Loeb Classical Library, London.

Sextus Empiricus: 1980, *Against the Logicians*. London, Loeb Classical Library, 1935.

Skyrms, Brian: 1980, *Causal Necessity*, Yale Univ. Press, New Haven.

Skyrms, Brian: 1980, 'Higher-Order Degrees of Belief', in D. H. Mellor (ed.), *Prospects for Pragmatism*, Cambridge Univ. Press, Cambridge.

Striker, Gisela: 1970, 'Sceptical strategies', in *Doubt and Dogmatism* See Schofield *et al*, 1980.

Whittle, Peter: 1970, *Probability*, Penguin, Baltimore.

Dept. of Philosophy
Princeton University
Princeton, New Jersey 08544
U.S.A.

HUGUES LEBLANC AND CHARLES G. MORGAN

PROBABILITY FUNCTIONS AND THEIR ASSUMPTION SETS – THE BINARY CASE

It is argued in Leblanc 1983b that statements are accorded probabilities *in light of assumptions* – or, as mathematicians often put it, *under conditions*.[1] It is further argued that each *singulary* probability function in Kolmogorov 1933 or, equivalently, Popper 1955 comes with (a set of) assumptions, to wit, those statements evaluating to 1 under P. With P a singulary probability function of the Kolmogorov-Popper sort and A a statement from a certain language L, Leblanc thus takes $P(A)$ to be the probability that P accords to A in light of the assumptions in $\{A : P(A) = 1\}$. His rationale for interpreting $P(A)$ in this manner is two-fold. He first contends that any assumption set in light of which a rational agent would accord probabilities *must* be deductively closed and, for convenience's sake, may be presumed consistent as well. He then establishes that a set S of statements of L is consistent and deductively closed if and only if there is a *singulary* probability function P for L such that $S = \{A : P(A) = 1\}$. The second result is called by Leblanc *The Fundamental Theorem on Assumption Sets, Case One*.

A like approach is taken here towards *binary* probability functions. We presume that *pairs of statements*, as well as single statements, are accorded probabilities *in light of assumptions*, and that each binary probability function P in Popper 1959 comes with (a set of) assumptions, to wit, those statements A of L such that – no matter the statement B of L – $P(A/B) = 1$. We thus take $P(A/B)$ to be the probability that P accords to A *in light of the assumptions in* $\{A : (\forall B)(P(A/B) = 1)\}$ *and in the light of B, B a stated* assumption which may but need not figure among the members of $\{A : (\forall B)(P(A/B) = 1)\}$, and when it does *not* figure among those *unstated* assumptions is operative only if flagged as in $P(A/B)$. Our rationale for interpreting $P(A/B)$ in this manner is two-fold. We take it as in Leblanc 1983b that *a rational agent would accord probabilities in light of all and only those sets of statements of L that are consistent and deductively closed*. And we establish (in Theorem 5 below) that *a set S of statements of L is consistent*

Synthese **60** (1984) 91–106.   0039–7857/84/0601–0091 $01.60
© 1984 *by D. Reidel Publishing Company*

and deductively closed if and only if there is a binary probability function P for L such that $S = \{A : (\forall B)(P(A/B) = 1)\}$ a result we call *The Fundamental Theorem on Assumption Sets, Case Two.*

When arguing that assumption sets must be deductively closed, Leblanc remarked that

> to be rational is to be alert to logical implications, those of what one knows, those of what one believes, and – in the present context – those of what one assumes. So, a rational agent – and it is for rational agents that the probability functions in studies such as this are intended – would own as an assumption any statement of L logically implied by his assumptions, and hence accord probabilities in light of *none but* deductively closed sets of statements of L.

Leblanc could see according probabilities in light of "contradictory" assumptions, so long as the probabilities accorded all equal 1. However, the constraints Kolmogorov and Popper place on *singulary* probability functions ruled that out. **C0**, one of the constraints Popper places on *binary* probability functions, rules it out here as well. The constraint, however, could be dropped without prejudice (other than editorial) to our results, a matter we take up on p. 100.[2]

The authors first thought of assumption sets in 1979–80 while outfitting intuitionistic logic with a "probabilistic" semantics, and they obtained (roughly) the present proof of the Fundamental theorem, Case Two, in the summer of 1980. Proof of Case One was found in early 1982 and is reported in Leblanc 1983b.[3]

* * *

The language L we work with is that in Leblanc 1983b. It has as its *primitive signs* $\aleph_0$ *atomic* statements, the two connectives ' $\sim$ ' and '&', and the two parentheses '(' and ')'; as its *formulas* all finite sequences of primitive signs of L; and as its *statements* (i) the atomic statements just mentioned, (ii) all formulas of the sort $\sim A$, where A is a statement of L, and (iii) all those of the sort (A & B), where each of A and B is a statement of L. We shall presume the statements of L to have been arranged in a fixed order, to be known as their *alphabetic order*; one way of doing that, due to Smullyan, is reported in Leblanc and Wisdom 1972. To abridge matters, we shall write (A ⊃ B) for $\sim (A \& \sim B)$; with A the alphabetically earliest statement of L, we shall write T for (A ⊃ A); and, when clarity permits, we shall drop outer parentheses.

L has as its *axioms* all statements of L of the sorts $A \supset (A \& A)$, $(A \& B) \supset A$, and $(A \supset B) \supset (\sim(B \& C) \supset \sim(C \& A))$, and as its one *rule of inference* Modus Ponens – with B called as in Quine *the ponential of A and $A \supset B$*. Where A is a statement and S a set of statements of L, a finite column of statements of L is counted a *proof of A from S* if (i) every entry in the column is a member of S, an axiom, or the ponential of two earlier entries in the column, and (ii) A is the last entry in the column; A is counted *provable from S* ($S \vdash A$, for short) if there exists a proof of A from S; and A is counted *provable* ($\vdash A$, for short) if $\phi \vdash A$. To define formally terms used above and congeners of these, a set S of statements of L is counted *consistent* if there is no statement A of L such that both $S \vdash A$ and $S \vdash \sim A$, *inconsistent* otherwise; *complete* if for each statement A of L either $S \vdash A$ or $S \vdash \sim A$, *incomplete* otherwise; *deductively closed* if S has among its members each statement of L provable from it; and *maximally consistent* if S is consistent, complete, and deductively closed. Further, we say that A is *consistent with S* if $S \cup \{A\}$ is consistent, and that S and S' are *deductively equivalent* if any statement of L provable from either of S and S' is provable from the other.

One more syntactic notion will be defined on p. 95, *the Lindenbaum extension $L(S)$ of a set S of statements of L*; and one more on p. 101 that of *a state-description of L*.

Turning to semantic (hence, probabilistic) matters, we understand by *a (binary) probability function for L* any function P that maps the pairs of statements of L into reals and meets the following seven constraints (issuing from constraints in Popper 1959):[4]

C0. $(\exists A)(\exists B)(P(A/B) \neq 1)$ (Existence)
C1. $0 \leq P(A/B)$ (Nonnegativity)
C2. $P(A/A) = 1$ (Reflexivity)
C3. If $(\exists C)(P(C/B) \neq 1)$, then
     $P(\sim A/B) = 1 - P(A/B)$ (Complementation)
C4. $P(A \& B/C) = P(A/B \& C) \times P(B/C)$ (Multiplication)
C5. $P(A \& B/C) \leq P(B \& A/C)$ (Commutation to the left)
C6. $P(A/B \& C) \leq P(A/C \& B)$ (Commutation to the right)

With P a (binary) probability function of L, we take a statement A of L to be *P-normal* if $(\exists B)(P(B/A) \neq 1)$,[5] otherwise to be *P-abnormal*; and, as indicated earlier, we take a set S of statements of L to constitute *the assumption set of P* if $S = \{A : (\forall B)(P(A/B) = 1)\}$. With S again a set of statements of L, we take S to constitute *a binary assumption set of L* – for short, *an assumption set of L* – if there exists a (binary) probability function for L of which S is the assumption set.[6] And, with S once more a set of statements and A a statement of L, we say that S *logically implies A in the probabilistic sense* if – no matter the probability function P for L – $(\forall B)(P(A/B) = 1)$ if $(\forall B)(P(C/B) = 1)$ for each member C of S, and that A *is logically true* (i.e., is a tautology) *in the probabilistic sense* if ϕ logically implies A in that sense. S thus logically implies A in the probabilistic sense if and only if A belongs to each assumption set of L of which S is a subset, and A is logically true in that sense if and only if A belongs to each assumption set of L, two results which speak for our account of an assumption set.

Proof that

(1) *If $S \vdash A$, then S logically implies A in the probabilistic sense,*

a probabilistic version of the *Strong Soundness Theorem for L*, will be found in Leblanc 1979 and (considerably improved) in Leblanc 1983b, as will be proof that

(2) *If S logically implies A in the probabilistic sense, then $S \vdash A$,*

a probabilistic version of the *Strong Completeness Theorem* for L. (1)–(2) legitimize the account of logical implication and – with ϕ as S – logical truth. We appeal to (1) in the proof of Theorem 1, and show on pp. 99–100 that (2) follows from Theorem 3(a).

$$* \quad * \quad *$$

Half of the Fundamental theorem on Assumption sets, Case Two, is readily proved:

THEOREM 1. *If there is a probability function P for L such that $S = \{A : (\forall B)(P(A/B) = 1)\}$, then S is deductively closed and consistent.*

Proof: Let P be an arbitrary probability function for L such that $S = \{A : (\forall B)(P(A/B) = 1)\}$. (1) Let A be an arbitrary statement of L such that $S \vdash A$. Then, by the Strong Soundness Theorem for L, A belongs to each assumption set of L of which S is a subset. But by hypothesis S is an assumption set of L, and S is of course a subset of itself. Hence $A \in S$. Hence S is deductively closed. (2) Suppose for reductio that S is inconsistent. Then, no matter the statement A of L, $S \vdash A$, hence by (1) $A \in S$, and hence $(\forall B)(P(A/B) = 1)$, against

Constraint **C0**. Hence S cannot be inconsistent.[7]

Proof of the converse of Theorem 1 calls for an extra notion and a few preliminary results.

The reader is doubtless familiar with the so-called *Lindenbaum extension* $L(S)$ of a set S of statements of L. We define $L(S)$ by means of (i)–(iii) below, and – using various Lindenbaum extensions of the sort $L(S \cup \{\cdot\})$ – we construct in Theorem 2 a binary function P_S which (a) invariably meets Constraints **C1–C6** and (b) when S is consistent, meets Constraint **C0** as well.

Let S be an arbitrary set of statements of L, and A_1, A_2, A_3, etc., be in alphabetic order the various statements of L.

(i) $L_0(S)$ is to be S itself,

(ii)   for each i from 1 on

$$L_i(S) = \begin{cases} L_{i-1}(S) \cup \{A_i\} \text{ if } L_{i-1}(S) \text{ is inconsistent or } L_{i-1}(S) \cup \{A_i\} \\ \text{is consistent} \\ L_{i-1}(S) \cup \{\sim A_i\} \text{ otherwise,}^8 \end{cases}$$

and

(iii) $L(S)$ is to be the union of $L_0(S)$, $L_1(S)$, $L_2(S)$, etc.

We shall take for granted various familiar facts about Lindenbaum extensions in general: S is a subset of $L(S)$, $L_i(S)$ ($i = 1, 2, 3, \ldots$) and $L(S)$ are consistent if S is, $L(S)$ is complete, and $L(S)$ is deductively closed. Further facts about these extensions are separately recorded in Lemmas 1 and 2, and one fact about Lindenbaum extensions of the sort $L(S \cup \{\cdot\})$ is recorded in Lemma 3.

LEMMA 1. *Let S be a set of statements of L; let A_1, A_2, $A_3, \ldots$, be in alphabetic order the various statements of L; and for each i from 1 on let B_i be A_i or $\sim A_i$.*

(a) *If S is inconsistent, then $L(S) = S \cup \{A_1, A_2, A_3, \ldots\}$;*

(b) *If S is consistent, then*

$$L_i(S) = L_{i-1}(S) \cup \{A_i\}$$

or

$$L_i(S) = L_{i-1}(S) \cup \{\sim A_i\}$$

according as A_i is consistent with $L_{i-1}(S)$ or $\sim A_i$ is;

(c) *Whether or not S is consistent,* (i) $L(S) = S \cup \{B_1, B_2, B_3, \ldots\}$ *and* (ii) *each member of S is one of* $B_1, B_2, B_3, \ldots$;
   (d) *If S is consistent and* $L(S) \vdash A$, *then A is consistent with each subset of* $L(S)$.

Clause (b) of the lemma hinges upon the fact that $L_{i-1}(S)$ is consistent if S is, and either $L_{i-1}(S) \cup \{A_i\}$ or $L_{i-1}(S) \cup \{\sim A_i\}$ is consistent if $L_{i-1}(S)$ is. (ii) in Clause (c) follows from (a) when S is inconsistent. In the contrary case let member C of S be the alphabetically i-th statement of L. Since $L_i(S)$ is consistent, $L_i(S)$ is sure to be $L_{i-1}(S) \cup \{C\}$ rather than $L_{i-1}(S) \cup \{\sim C\}$, and hence C is sure to be B_i. And Clause (d) hinges upon the fact that $A \in L(S)$ if $L(S) \vdash A$, and each subset of $L(S)$ is sure to be consistent if S is. So, under the present circumstances each subset $S' \cup \{A\}$ of $L(S)$ is sure to be consistent, and hence A is sure to be consistent with each subset S' of $L(S)$.

LEMMA 2. *If S and S' are deductively equivalent, then* $L(S) = L(S')$.
   Proof: Let S and S' be deductively equivalent. When S is inconsistent, so of course is S', and hence by Lemma 1(a) $L(S) = L(S')$. So, suppose S is consistent, and let

$$L(S) = S \cup \{B_1, B_2, B_3, \ldots\},$$

where B_1, B_2, B_3, etc., are as in Lemma 1. (1) Since S and S' are deductively equivalent, B_1 is consistent with S' as well as with S, and hence by Lemma 1(b) $L_1(S') = S' \cup \{B_1\}$. (2) Since as a result $L_1(S)$ and $L_1(S')$ are deductively equivalent, B_2 is consistent with $L_1(S')$ as well as with $L_1(S)$, and hence by Lemma 1(b) again $L_2(S') = S' \cup \{B_1, B_1\}$. And so on. Hence for each i from 1 on $L_i(S') = S \cup \{B_1, B_2, \ldots B_i\}$, and hence

$$L(S') = S' \cup \{B_1, B_2, B_3, \ldots\}.$$

But by Lemma 1(c) each member of S is among B_1, B_2, B_3, etc., as is each member of S'. Hence $L(S) = L(S')$. Hence Lemma 2.

LEMMA 3. *If* $L(S \cup \{A\}) \vdash B$, *then* $L(S \cup \{A\}) = L(S \cup \{B \, \& \, A\})$.
   Proof: When $S \cup \{A\}$ is inconsistent, so is $S \cup \{B \, \& \, A\}$, and hence by Lemma 1(a) $L(S \cup \{A\}) = L(S \cup \{B \, \& \, A\})$. So, suppose $S \cup \{A\}$ is consistent, and for arbitrary i equal to or larger than 1 let

$$L_i(S \cup \{B \& A\}) = S \cup \{B \& A\} \cup \{C_1, C_2, \ldots C_i\}.$$

Since $L_i(S \cup \{A\})$ is sure to be consistent if $S \cup \{A\}$ is, C_1 is consistent with $S \cup \{B \& A\}$, C_2 with $S \cup \{B \& A\} \cup \{C_1\}$, etc. But, if so, then C_1 is consistent with $S \cup \{A\}$, C_2 with $S \cup \{A\} \cup \{C_1\}$, etc., and hence by Lemma 1(b)

$$L_i(S \cup \{A\}) = S \cup \{A\} \cup \{C_1, C_2, \ldots, C_i\}.$$

Suppose then that

$$L_i(S \cup \{A\}) = S \cup \{A\} \cup \{C_1', C_2', \ldots, C_i'\},$$

and suppose $L(S \cup \{A\}) \vdash B$. Then by Lemma 1(d) B is consistent with $S \cup \{A\} \cup \{C_1'\}$, and hence C_1' is consistent with $S \cup \{B \& A\}$. Hence by the same reasoning C_2' is consistent with $S \cup \{B \& A\} \cup \{C_1'\}$. And so on. Hence

$$L_i(S \cup \{B \& A\}) = S \cup \{B \& A\} \cup \{C_1', C_2', \ldots, C_i'\}.$$

Hence

$$L(S \cup \{A\}) = S \cup \{A\} \cup S'$$

and

$$L(S \cup \{B \& A\}) = S \cup \{B \& A\} \cup S'$$

for some common set S' of statements of L. But by Lemma 1(c) each of A and $B \& A$ is sure to be a member of S'. Hence $L(S \cup \{A\}) = L(S \cup \{B \& A\})$. Hence Lemma 3.

With Lemmas 2 and 3 at hand we show that any set S of statements of L generates a function for L meeting Constraints **C1–C6**; and, when S is consistent, one meeting Constraint **C0** as well. The result will readily yield the converse of Theorem 1.

THEOREM 2. *Let S be a set of statements of L, and P_S be the binary function such that, for any two statements A and B of L,*

$$P_S(A/B) = \begin{cases} 1 & \text{if } L(S \cup \{B\}) \vdash A \\ 0 & \text{otherwise.} \end{cases}$$

Then:

(a) *P_S meets Constraints **C1–C6**;*

(b) *If S is consistent, then P_S meets Constraint **C0** as well.*

Proof: (a) That P_S meets Constraints **C1–C6** can be shown as follows.
Constraint **C1**: $0 \le P_S(A/B)$ by the very construction of P_S.
Constraint **C2**: $L(S \cup \{A\}) \vdash A$. Hence $P_S(A/A) = 1$.
Constraint **C3**: Suppose $(\exists C)(P_S(C/B) \ne 1)$. Then $(\exists C)(L(S \cup \{B\}) \nvdash C)$, and hence $L(S \cup \{B\})$ is consistent. Suppose next that $P_S(A/B) = 1$, in which case $L(S \cup \{B\}) \vdash A$. Then $L(S \cup \{B\}) \nvdash \sim A$ by the consistency of $L(S \cup \{B\})$, and hence $P_S(\sim A/B) = 0$. Suppose finally that $P_S(A/B) = 0$. Then $L(S \cup \{B\}) \nvdash A$, hence by the completeness of $L(S \cup \{B\})$, $L(S \cup \{B\}) \vdash \sim A$, and hence $P_S(\sim A/B) = 1$. Hence, if $(\exists C)(P_S(C/B) \ne 1)$, then $P_S(\sim A/B) = 1 - P_S(A/B)$.
Constraint **C4**: Suppose *first* that $P_S(A \& B/C) = 1$. Then $L(S \cup \{C\}) \vdash A \& B$, and hence both $L(S \cup \{C\}) \vdash A$ and $L(S \cup \{C\}) \vdash B$. But, if $L(S \cup \{C\}) \vdash B$, then by Lemma 3 $L(S \cup \{C\}) = L(S \cup \{B \& C\})$, and hence $L(S \cup \{B \& C\}) \vdash A$. Hence both $P_S(A/B \& C)$ and $P_S(B/C)$ equal 1, and hence $P_S(A \& B/C) = P_S(A/B \& C) \times P_S(B/C)$. Suppose *next* that $P_S(A \& B/C) = 0$. Then $L(S \cup \{C\}) \nvdash A \& B$, hence either $L(S \cup \{C\}) \nvdash A$ or $L(S \cup \{C\}) \nvdash B$ (or both), and hence either $P_S(A/C) = 0$ or $P_S(B/C) = 0$ (or both). Now suppose $P_S(B/C) \ne 0$, in which case $P_S(A/C) = 0$. Then $P_S(B/C) = 1$, hence $L(S \cup \{C\}) \vdash B$, hence by Lemma 3 $L(S \cup \{C\}) = L(S \cup \{B \& C\})$, hence $L(S \cup \{C\}) \vdash A$ if and only if $L(S \cup \{B \& C\}) \vdash A$, hence $P_S(A/C) = P_S(A/B \& C)$, and hence $P_S(A/B \& C) = 0$. Hence, if $P_S(B/C) \ne 0$, then $P_S(A/B \& C) = 0$. Hence $P_S(B/C) = 0$ or $P_S(A/B \& C) = 0$. Hence again $P_S(A \& B/C) = P_S(A/B \& C) \times P_S(B/C)$.
Constraint **C5**: $L(S \cup \{C\}) \vdash A \& B$ if and only if $L(S \cup \{C\}) \vdash B \& A$, hence $P_S(A \& B/C) = P_S(B \& A/C)$, and hence $P_S(A \& B/C) \le P_S(B \& A/C)$.
Constraint **C6**: Both when consistent and when not, $S \cup \{B \& C\}$ and $S \cup \{C \& B\}$ are deductively equivalent. Hence by Lemma 2 $L(S \cup \{B \& C\}) = L(S \cup \{C \& B\})$, hence $L(S \cup \{B \& C\}) \vdash A$ if and only if $L(S \cup \{C \& B\}) \vdash A$, hence $P_S(A/B \& C) = P_S(A/C \& B)$, and hence $P_S(A/B \& C) \le P_S(A/C \& B)$.

(b) Suppose S is consistent. Then $S \cup \{T\}$ is consistent, hence so is $L(S \cup \{T\})$, hence $L(S \cup \{T\}) \nvdash \sim T$, hence $P_S(\sim T/T) = 0$, hence $(\exists A)(\exists B)(P_S(A/B) \ne 1)$, and hence P_S meets Constraint **C0** as well as Constraints **C1-C6**. Hence (b).

THEOREM 3. (a) *If a set S of statements of L is consistent, then there is a probability function P for L such that* $\{A : S \vdash A\} = \{A : (\forall B)(P(A/B) = 1)\}$. (b) *If S is consistent and deductively closed, then there is a probability function P for L such that* $S = \{A : (\forall B)(P(A/B) = 1)\}$.

Proof: Let S be consistent.

(a) Let P_S be the function defined in the preamble of Theorem 2, and suppose *first* that $S \vdash A$. Then $(\forall B)(L(S \cup \{B\}) \vdash A)$, and hence $(\forall B)$ $(P_S(A/B) = 1)$ by the definition of P_S in Theorem 2. Suppose *next* that $S \nvdash A$. Then $S \cup \{\sim A\}$ is consistent, and hence so is $L(S \cup \{\sim A\})$. But $L(S \cup \{\sim A\}) \vdash \sim A$. Hence $L(S \cup \{\sim A\}) \nvdash A$, hence $P_S(A/\sim A) = 0$ by the definition of P_S, and hence $(\exists B)(P_S(A/B) \neq 1)$. Hence $S \vdash A$ if and only if $(\forall B)(P_S(A/B) = 1)$. Hence (a).

(b) Suppose S is deductively closed. Then $S = \{A : S \vdash A\}$. Hence (b) by (a).

Hence:

THEOREM 4. *Any set of statements of L that is consistent and deductively closed is the assumption set of a probability function for L.*

Hence Case Two of the Fundamental Theorem on Assumption Sets:

THEOREM 5. *A set of statements of L is the assumption set of a* (binary) *probability function for L if and only if it is consistent and deductively closed.*

As the reader may verify, $S \vdash A$ if and only if A belongs to each consistent and deductively closed set of L of which S is a subset. Theorem 5 thus guarantees that

(3) *If $S \vdash A$, then A belongs to each assumption set of L of which S is a subset,*

and

(4) *If A belongs to each assumption set of L of which S is a subset, then $S \vdash A$,*

and hence – given the account of logical implication on p. 94 – the theorem yields each of (1) and (2) on that page.

That Theorem 5 yields (1) signifies little since we called on that version of the Soundness Theorem for L when proving Theorem 1. That Theorem 5 yields (2) is of more interest, as indeed may be the following proof of (2) by means of just Theorem 3(a). Suppose that $S \nvdash A$, in which case $S \cup \{\sim A\}$ is sure to be consistent. Then by Theorem

3(a) there exists a probability function for L, call it $P_{S \cup \{\sim A\}}$, such that (i) $(\forall C)(P_{S \cup \{\sim A\}}(B/C) = 1)$ for each statement B of L that belongs to S and hence is provable from S, and (ii) $(\forall C)(P_{S \cup \{\sim A\}}(\sim A/C) = 1)$ and hence $P_{S \cup \{\sim A\}}(\sim A/T) = 1$. But, as a brief argument using Constraint **C0** would show, T is P-normal, this for each probability function P for L.[9] Hence by Constraint **C3** $P_{S \cup \{\sim A\}}(A/T) = 0 \neq 1$. Hence there exists a probability function P for L such that $(\forall C)(P(B/C) = 1)$ for each member B of S and yet $(\exists C)(P(A/C) \neq 1)$. Hence, by Contraposition, A is sure – if logically implied by S in the probabilistic sense – to be provable from S.

Returning to a point raised on p. 92, suppose that Constraint **C0** were dropped. Theorems 1 and 2(a) would then guarantee that a set of statements of L constitutes an assumption set of L if and only if it is deductively closed. The result is easily accommodated here. We just pointed out that $S \vdash A$ if and only if A belongs to each consistent and deductively closed set of statements of L with S as a subset. However, it can also be shown that $S \vdash A$ if and only if A belongs to each deductively closed set of statements of L – be the set consistent or not – with S as a subset. So, with **C0** dropped from the list of constraints on p. 93, A would still be provable from S if and only if it belongs to each assumption set of L with S as a subset. In view of this result and others, one might willingly jettison **C0**.[10]

* * *

The assumption sets of L can be sorted as follows:

Group One, to consist of $\{A: \vdash A\}$, a set included in all the assumption sets of L,
Group Two, to consist of the remaining assumption sets of L that are *incomplete*, and
Group Three, to consist of all the assumption sets of L that are *complete* – hence, to consist of the "maximally consistent" sets of statements of L.

The set in Group One is the "smallest" assumption set of L: it is the only assumption set of L no proper subset of which constitutes an assumption set of L. The sets in Group Three, on the other hand, are the "largest" assumption sets of L as befits *maximally* consistent sets: they are the only assumption sets of L no proper superset of which constitutes an assumption set of L.

The probability functions for L with $\{A : \vdash A\}$ as their assumption set are called in Leblanc and van Fraassen 1979 *Carnap's probability functions* for L.[11] They are the probability functions for L that meet the following constraint (due to van Fraassen and to Fine, independently):

C7. No matter the state-description C of L, $P(\sim C/C) = 0$,

a *state-description* of L being a conjunction of the sort $(\ldots (\pm A_1 \ \& \ \pm A_2) \ \& \ldots) \ \& \ \pm A_n$, where $n \geq 1$, $A_1, A_2, \ldots, A_n$ are in alphabetic order the first n atomic statements of L, and for each i from 1 through n, $\pm A_i$ is A_i or $\sim A_i$.[12]

LEMMA 4. *Let C be an arbitrary state-description of L.*

(a) *If $P(\sim C/C) = 0$, then – no matter the statement A of L – $(\forall B)(P(A/B) = 1)$ if and only if $\vdash A$.*

(b) *If – no matter the statement A of L – $(\forall B)(P(A/B) = 1)$ if and only if $\vdash A$, then $P(\sim C/C) = 0$.*

Proof: (a) Let $P(\sim C/C) = 0$, and hence $(\exists B)(P(B/C) \neq 1)$; and suppose first that $\vdash A$. Then $(\forall B)(P(A/B) = 1)$ by the Soundness Theorem for L. Suppose next that $\nvdash A$. Then there is sure to be a state-description C of L such that $\vdash C \supset \sim A$, hence by a familiar result such that $P(\sim A/C) = 1$,[13] and hence by **C3** (and $(\exists B)(P(B/C) \neq 1)$) such that $P(A/C) = 0 \neq 1$. Hence $(\exists B)(P(A/B) \neq 1)$. Hence $(\forall B)(P(A/B) = 1)$ if and only if $\vdash A$. Hence (a).[14]

(b) Let $(\forall B)(P(A/B) = 1)$ if and only if $\vdash A$, this for any statement A of L. Since $\nvdash \sim C$, $(\exists B)(P(\sim C/B) \neq 1)$, and hence by a familiar result $P(\sim C/C) = 0$.[15] Hence (b).

Hence:

THEOREM 6. $\{A : \vdash A\}$ *is the assumption set of all and only those probability functions for L known as Carnap's probability functions.*[16]

The probability function P_S in Theorem 2 being 2-valued, each assumption set of L – i.e., $\{A : \vdash A\}$, each set in Group Two, and each one in Group Three – is the assumption set of at least one 2-valued probability function for L. $\{A : \vdash A\}$ and the assumption sets in Group Two are also the assumption sets of probability functions boasting more than 2 values. Not so, however, the sets in Group Three. All probability functions with these sets as assumption sets are indeed 2-valued (Theorem 7(b)).

Further, $\{A : \vdash A\}$ is the assumption set of $2^{\aleph_0}$ probability functions, as is each of the $2^{\aleph_0}$ sets in Group Two. Each of the $2^{\aleph_0}$ sets in Group Three, by contrast, is the assumption set of exactly one probability function (Theorem 8(a)). However, since $2^{\aleph_0} \times 2^{\aleph_0}$ equals $2^{\aleph_0}$, the probability functions with a set from Group Two as their assumption set are $2^{\aleph_0}$ in number, as are those with $\{A : \vdash A\}$ or with a set from Group Two as their assumption set.

LEMMA 5. *Let S be a maximally consistent set of statements of L, and P be any probability function for L such that* $S = \{A : (\forall B)(P(A/B) = 1)\}$. *Then, for any statements A and B of L, $P(A/B)$ equals 1 or 0 according as $B \supset A$ belongs to S or not.*

Proof: Let A and B be arbitrary statements of L, and suppose *first* that B is P-abnormal. Then by definition $P(A/B) = 1$, and hence $P(A/B) = 1$ if $B \supset A \in S$. Suppose *then* that B is P-normal and $B \supset A \in S$. Then $\sim B \in S$ or $A \in S$; hence, by the hypothesis on S and that on P, $(\forall C)(P(\sim B/C) = 1)$ or $(\forall C)(P(A/C) = 1)$; and hence $P(\sim B/B) = 1$ or $P(A/B) = 1$. But $P(\sim B/B) = 0 \neq 1$ by **C2, C3**, and the P-normality of B. Hence $P(A/B) = 1$. Suppose *finally* that B is P-normal but $B \supset A \notin S$. Then $\sim A \in S$ by the hypothesis on S, hence $(\forall C)(P(\sim A/C) = 1)$ by the hypothesis on S and that on P, hence $P(\sim A/B) = 1$, and hence $P(A/B) = 0$ by **C3** and the P-normality of B. Hence Lemma 5.

Hence:

THEOREM 7. *Let S be a maximally consistent set of L. Then:*

(a) *S is the assumption set of exactly one probability function for L;*

(b) *The one probability function for L of which S is the assumption set is 2-valued.*

Proof: (a) Suppose $S = \{A : (\forall B)(P_1(A/B) = 1)\}$ and $S = \{A : (\forall B)(P_2(A/B) = 1)\}$. Then by Lemma 5, $P_1(A/B) = P_2(A/B)$ for any statements A and B of L. (b) By (a) and Lemma 5.

$*$ $*$ $*$

It is because of Theorem 5, to wit:

(5) $(\exists P)(S = \{A : (\forall B)(P(A/B) = 1)\})$ *if and only if S is consistent and deductively closed,*

and our feeling that all and only consistent and deductively closed sets of statements of L qualify as assumption sets, that we appointed $\{A : (\forall B)(P(A/B) = 1)\}$ the assumption set of P. Alternatively, we could have required of the assumption set S of P that

(6) $S = \{A : (\forall B)(\forall C)(P(A \& B/C) = P(B/C))\}$

and

(7) $S = \{A : (\forall B)(\forall C)(P(B/A \& C) = P(B/C))\}$.

Showing that S meets desiderata (6)–(7) *if and only if*

(8) $S = \{A : (\forall B)(P(A/B) = 1)\}$

would have been an easy task, and Theorem 5 would now assure us that a set of statements of L counts as an assumption set if and only if it is consistent and deductively closed.

The route is a longer one, but going about things thusly would effectively rule out

$$\{A : P(A/T) = 1\}$$

from consideration as the assumption set of P. $\{A : P(A/T) = 1\}$ does have credentials. It can indeed by shown, as Leblanc noted in early 1982, that

(9) $S \vdash A$ *if and only if – no matter and probability function P for L – P(A/T) = 1 if P(B/T) = 1 for each member B of S,*[17]

and

(10) $(\exists P)(S = \{A : P(A/T) = 1\})$ *if and only if S is consistent and and deductively closed.*

Because of (9) one might take S *to logically imply A in the probabilistic sense* if – no matter the probability function P for L – $P(A/T) = 1$ if $P(B/T) = 1$ for each member B of S. Despite (10), however, one should not understand $\{A : P(A/T) = 1\}$ as the assumption set of L: it does not follow form **C1–C6** that

(11)   *If* $P(A/T) = 1$, *then* $(\forall B)(\forall C)(P(B/A \& C) = P(B/C))$.

For proof, assign 2 to A, 0 to B, and 1 to C. According to the matrices on p. 338 of Popper 1959, $P(A/T)$ will then equal 1, and yet $P(B/A \& C)$ will equal 1 while $P(C/B)$ equals 0.

Interestingly enough, $\{A : (\forall B)(P(A/B) = 1)\}$ and $\{A : P(A/T) = 1\}$ are the same, Morgan recently noted, when P is what we acknowledge in Morgan and Leblanc 1983a as an intuitionistic probability function for L. But intuitionistic logic is not our present concern.[18,19]

NOTES

[1] All pertinent texts are listed in the References. We refer to them in the body of the paper and in the notes by author's name and date of publication.

[2] Theorem 5 holds with $(\exists P)(S = \{A : P(A/T) = 1\})$ in place of $(\exists P)(S = \{A : (\forall B) (P(A/B) = 1)\})$. But, as we shall see on pp. 13–14, $\{A : P(A/T) = 1\}$ is unsuitable as the assumption set of P.

[3] A paper from which we freely borrow when covering matters already treated there.

[4] See p. 349. That Popper's constraints **A1–A3**, **B1–B2**, and **C** are equivalent to **C0–C6** is shown in Harper, Leblanc, and van Fraassen 1983 and in Leblanc 1981. In earlier writings of Leblanc's **C5–C6** appeared as equalities; we use inequalities here to preserve consistency with Popper 1955, Leblanc 1983a, and Leblanc 1983b; in those texts the singular counterpart of **C5** runs $P(A \& B) \leq P(B \& A)$ (rather than $P(A \& B) = P(B \& A)$).

[5] Given this definition **C3** may of course be made to read: "If B is P-normal, then $P(\sim A/B) = 1 - P(A/B)$."

[6] From now on we shall often drop the (already parenthesized) 'binary'.

[7] The proof is an adaption to the binary case of the proof of Theorem 1 in Leblanc 1983b.

[8] $L(S)$ is commonly defined only for consistent S, in which case $L_i(S)$ is taken to be $L_{i-1}(S) \cup \{A_i\}$ when $L_{i-1}(S) \cup \{A_i\}$ is consistent, otherwise to be $L_{i-1}(S) \cup \{\sim A_i\}$.

[9] See the proof of T5.33(c) in Leblanc 1983a.

[10] Only one set of statements of L is both inconsistent and deductively closed: the set of all the statements of L. The probability function with that set as its assumption set would of course be the function P such that $(\forall A)(\forall B)(P(A/B) = 1)$. See Leblanc 1983b, pp. 381 and 395, for more on requiring assumption sets to be consistent and for ways of lifting that requirement in the singular case.

[11] Also *Popper's probability functions in the narrow sense*, an unfortunate appellation which we hope will not gain currency.

[12] In the presence of **C7**, **C0** becomes of course redundant.

[13] The result trivially holds true when C is P-abnormal. So suppose C is P-normal, and suppose $\vdash C \supset \sim A$ (i.e., $\vdash \sim(C \& A)$). Then by the Soundness Theorem for L $P(\sim(C \& A)/C) = 1$, hence by **C3** $P(C \& A/C) = 0$, and hence by **C4** $P(C/A \& C) \times P(A/C) = 0$. But, as we establish a few lines hence, $P(C/A \& C) = 1$. Hence $P(A/C) = 0$, and hence by **C3** and the P-normality of C $P(\sim A/C) = 1$. For proof that

$P(C/A \& C) = 1$, note that by **C2** $P(A \& C/A \& C) = 1$ and hence by **C4** $P(A/C \& (A \& C)) \times P(C/A \& C) = 1$. But by **C1** and **C3** any probability lies in the interval $[0, 1]$, and hence each of two probabilities must equal 1 if their product does. Hence $P(C/A \& C)$ has to equal 1.

[14] The first correct proof of (a), an embarrassingly long one, is in Leblanc and van Fraassen 1979.

[15] We prove that if $P(\sim C/C) \neq 0$, then $(\forall B)(P(\sim C/B) = 1)$. The result trivially holds true when B is P-abnormal. So suppose B is P-normal (Hypothesis One) and $P(\sim C/C) \neq 0$ (Hypothesis Two). By Hypothesis One and **C2–C3**, $P(\sim B/B) = 0$, hence by **C4** $P(C \& \sim B/B) = 0$, hence by **C5** $P(\sim B \& C/B) = 0$, and hence by **C4** $P(\sim B/C \& B) \times P(C/B) = 0$. But by Hypothesis Two and **C2–C3**, $(\forall B)(P(B/C) = 1)$, hence $P(\sim B \& B/C) = 1$, hence by **C4** and the result obtained at the close of note 13, $P(\sim B/B \& C) = 1$, and hence by **C6** $P(\sim B/C \& B) = 1$. Hence $P(C/B) = 0$, and hence by **C3** (and the P-normality of B) $P(\sim C/B) = 1$.

[16] For further information on Carnap's probability functions, see Leblanc and van Fraassen 1979, and Harper, Leblanc, and van Fraassen 1983.

[17] The proofs of the Soundness and Completeness Theorems in Leblanc 1983b are easily edited to yield (9). As for (10), establish first that the results of putting $P(\cdot/T)$ for $P(\cdot)$ in Constraints **C1–C6** of Leblanc 1983b follow from Constraints **C0–C6** in this paper. Putting $P(\cdot/T)$ everywhere for $P(\cdot)$ in the proof of Theorem 4 in Leblanc 1983b will then yield (10).

[18] Proving Theorem 5 for a language L with quantifiers as well as connectives is our next order of business. Bas van Fraassen reported in October 1982 that such a proof can be retrieved from his 1982 paper. However, the constraints he places there on binary probability functions are more restrictive than the ones commonly used.

[19] The paper is an elaboration of part of Leblanc's talk at the Conference on Foundations. While working on these matters Leblanc held a research grant from the National Science Foundation (Grant SES 8007179) and was on partial research leave from Temple University. Thanks are due to Tom McGinness, Muffy E. Siegel, and Bas van Fraassen for reading an earlier draft of the paper.

REFERENCES

Carnap, R.: 1950, *Logical Foundations of Probability*, Univ. of Chicago Press, Chicago.
Carnap, R.: 1952, *The Continuum of Inductive Methods*, Univ. of Chicago Press, Chicago.
Harper, W. L., H. Leblanc, and B. C. van Fraassen: 1983, 'On Characterizing Popper and Carnap Probability Functions', in H. Leblanc, R. D. Gumb, and R. Stern (eds.), *Essays in Epistemology and Semantics*, Haven Publishing, New York, pp. 117–139.
Kolmogorov, A. N.: 1933, *Grundbegriffe der Wahrscheinlichkeitsrechnung*, Berlin.
Leblanc, H.: 1979, 'Probabilistic Semantics for First-Order Logic', *Zeitschrift für Mathematische Logik und Grundlagen der Mathematik* **25**, 497–509.
Leblanc, H.: 1981, 'What Price Substitutivity? A Note on Probability Theory', *Philosophy of Science* **48**, 317–322.

Leblanc, H.: 1983a: 'Alternatives to Standard First-Order Semantics', in D. M. Gabbay and F. Guenthner (eds.), *Handbook of Philosophical Logic*, D. Reidel, Dordrecht, pp. 189–274.

Leblanc, H.: 1983b: 'Probability Functions and Their Assumption Sets: The Singulary Case', *Journal of Philosophical Logic* **12**, 379–402.

Leblanc, H. and B. C. van Fraassen: 1979: 'On Carnap and Popper Probability Functions', *The Journal of Symbolic Logic* **44**, 369–373.

Leblanc, H. and W. A. Wisdom: 1972. *Deductive Logic*, Allyn & Bacon, Inc., Boston.

Morgan, C. G. and H. Leblanc: 1983a: 'Probabilistic Semantics for Intuitionistic Logic', *Notre Dame Journal of Formal Logic* **24**, 161–180.

Morgan, C. G. and H. Leblanc: 1983b: 'Probability Theory, Intuitionism, Semantics, and the Dutch Book Argument', *Notre Dame Journal of Formal Logic* **24**, 289–304.

Popper, K. R.: 1955: 'Two Autonomous Axiom Systems for the Calculus of Probabilities', *The British Journal for the Philosophy of Science* **6**, 51–57, 176, 351.

Popper, K. R.: 1959, *The Logic of Scientific Discovery*, Basic Books, Inc., New York.

Quine, W. V.: 1940, *Mathematical Logic*, W. W. Norton & Co., New York.

Smullyan, R. N.: 1968, *First-Order Logic*, Springer-Verlag, New York.

van Fraassen, B. C.: 1982: 'Probabilistic Semantics Objectified: II. Implications in Probabilistic Model Sets', *Journal of Philosophical Logic* **10**, 495–510.

Dept. of Philosophy
Temple University
Philadelphia, PA19122
U.S.A.

GILBERT HARMAN

LOGIC AND REASONING

1. THE PROBLEM

Should we think of logic as a science like physics and chemistry, but more abstract and with a wider application? Or should we think of logic as having a special role in reasoning, a role that is not simply a consequence of its wider application? This is a difficult issue and I for one am unsure how to resolve it. In this paper I will try to say why certain answers are unsatisfactory.

The problem may be hard to appreciate. There is a tendency to identify reasoning with proof or argument in accordance with rules of logic. Given that identification, logic obviously has a special role to play in reasoning. But the identification is mistaken. Reasoning is not argument or proof. It is a procedure for revising one's beliefs, for changing one's view. (Reasoning also effects one's plans, intentions, desires, hopes, and so forth, but I am going to ignore that and concentrate on beliefs.) Reasoning often leads one to acquire new beliefs on the basis of one's old beliefs, but it also often leads one to abandon some old beliefs as well. There is subtracting as well as adding. The question, then, is whether logic has a special role to play in this procedure of belief revision.

Now logicians often speak of "rules of inference", by which they mean certain patterns of implication, such as *modus ponens*, which is sometimes formulated as follows: "From p and *if p then q*, infer q." Philosophers like Kneale, Dummett, and Hacking take this way of talking very seriously. They say it is a fundamental error to think of basic logical principles as axioms rather than as rules of inference.[1]

What do they mean? Logical principles are not directly rules of *belief revision*. They are not particularly about belief at all. For example, *modus ponens* does not say that , if one believes p and also believes *if p then q*, one may also believe q. Nor are there any principles of belief revision that directly correspond to logical principles like *modus ponens*. Logical principles hold universally, without exception, whereas

Synthese **60** (1984) 107–127.   0039–7857/84/0601–0107 $02.10

the corresponding principles of belief revision would be at best *prima facie* principles, which do not always hold. It is not always true that, if one believes *p* and believes *if p then q*, one may infer *q*. The proposition *q* may be absurd or otherwise unacceptable in the light of one's other beliefs, so that one should give up either one's belief in *p* or one's belief in *if p then q* rather than believe *q*. And, even if *q* is not absurd and is not in conflict with one's other beliefs, there may simply be no point to adding it to one's beliefs. The mind is finite. One does not want to clutter it with trivialities. It would be irrational to fill one's memory with as many as possible of the logical consequences of one's beliefs. That would be a terrible waste of time, leaving no room for other things.

I assume here a distinction between explicit and implicit beliefs. Given one's explicit beliefs, there are many other things one can be said to believe implicitly. These include things that are obviously implied by one's explicit beliefs. (They may also include things implicit in the *believing* of one's explicit beliefs, but I will ignore that here.[2]) There is a sense in which one can believe indefinitely many different things with room to spare. For example, one believes that $10,001 + 1 = 10,002$, that $10,002 + 1 = 10,003$, and so on. These things are obviously implied by one's explicit beliefs. I assume that there is a limit to what one can believe explicitly and that principles of inference are principles about the revision of explicit beliefs. Considerations of clutter-avoidance rule out cluttering one's mind with trivial *explicit* beliefs; but large numbers of implicit beliefs do not by themselves produce clutter.

The point then is that, whereas logical principles like *modus ponens* are exceptionless, corresponding rules of inference are not. Sometimes one should abandon a premise rather than accept a conclusion that follows logically from what one believes. And, perhaps more importantly, there is the worry about clutter.

It might be suggested that logical principles correspond to principles of reasoning saying what one should *not* believe. In this view the connection between logic and reasoning would be mediated by the rule, "Avoid inconsistency!" But even the rule "Avoid inconsistency!" has exceptions, if it requires one not to believe things one knows to be jointly inconsistent. On discovering one has inconsistent beliefs, one might not see any easy way to modify one's beliefs so as to avoid the inconsistency, and one may not have the time or ability to figure out the best response. In that case, one should (at least sometimes) simply acquiesce in the contradiction while trying to keep it fairly isolated. I would think this is the proper attitude for most ordinary people to take toward many paradoxical arguments.

Furthermore, a rational fallible person ought to believe that at least one of his or her beliefs is false. But then not all of his or her beliefs can be true, since, if all of the other beliefs are true, this last one will be false. So in this sense a rational person's beliefs are inconsistent. It can be proved they cannot all be true together.

Still, it might be said, there is some force to the principle "Avoid inconsistency!" even if this is only a prima facie or defeasible principle which does not hold universally. It holds "other things being equal". Why isn't that enough to give logic a special role in reasoning?

A possible reply is that this is enough only if there are not similar prima facie principles for physics, chemistry, and other sciences. But it would seem there are such principles. Prima facie, one should not continue to believe things one knows cannot all be true, whether this impossibility is logical, physical, chemical, mathematical, or geological.

On the other hand, it may be that, if you know certain beliefs cannot all be true, where this is a physical impossibility, then you must know that those beliefs are incompatible with your beliefs about physics. This would mean that the basic principle can after all be taken to be "Avoid inconsistency!" and logic could play a special role in reasoning.

I am not sure about this. Perhaps logic can play a special role in reasoning *via* inconsistency. If so, that is because of the abstractness and universal applicability of logic. Perhaps we can even define logic as the theory of inconsistency in the sense that the principles of logic are the minimum principles needed to convert all cases in which one knows certain beliefs cannot be true into cases in which those beliefs are logically incompatible with other beliefs one has. Or perhaps we can show that logic defined in some other way also has this property. But this is not the main issue I am concerned with.

I want to put aside this question about inconsistency for a moment, anyway, in order to consider whether logic should play any further role in reasoning. The question I want to consider is whether, in addition (perhaps) to playing a special role in reasoning in telling one what not to believe, logic (also) plays a special role in reasoning in telling one what one may believe.

One possibility, of course, is that logic has no such (further) special role to play. In this view logic is merely a body of truths, a science like physics or chemistry, but with a more abstract subject matter and therefore a more general application.

This is an extreme view that no one seems to hold in an unqualified way, which is surprising, since the view seems to be quite viable. Frege may seem to take the extreme view when he says the laws of logic are laws of truth and since he attacks "psychologism"; but he also says the laws of logic "prescribe universally the way in which one ought to think if one is to think at all",[3] which is to reject the extreme view. Similarly, Quine may seem to advocate the extreme view when he says logic is a science of truths.[4] But he also sees a special connection between logic and inference when he says one needs logic to get to certain conclusions from certain premises.[5] As far as I have been able to determine, other philosophers who may seem at one place to put forward the extreme view that logic is a science, a body of truths, go on some place else to say that logic has a special role to play in reasoning. I am not sure why I cannot find anyone who has unequivocally endorsed the extreme view.

It might be suggested that Lewis Carroll's story, 'What the Tortoise Said to Achilles', shows that logic cannot be treated as merely a body of truths.[6] But that is not so.

In the story, Achilles tries to get the tortoise to accept a conclusion Z. The tortoise accepts A and B but refuses to accept Z. Achilles argues that the tortoise must accept Z since he must agree that, if A and B then Z. The tortoise, being accomodating, agrees to accept C, namely, *if A and B then Z*. The tortoise also continues to accept A and B but still refuses to accept Z. Achilles then says the tortoise *must* now accept Z since this is required by logic. The tortoise agrees to accept anything required by logic and therefore accepts D, namely, *if A and B and C then Z*, but *still* refuses to accept Z. The story continues in this vein with Achilles getting the tortoise to accept more and more logical principles, E, F, G, and so on, without being able to get the tortoise to draw the final conclusion Z.

It might be said that the absurdity of the story lies in confusing a rule of inference with a premise. And, it might be said, the moral of the story is that logic is not just a body of truths but includes rules of inference as well. However, the story shows no such thing.

For one thing, there is a sense in which it is not enough for the tortoise to accept a rule of inference. Suppose the tortosie agrees to accept the rule R, "From A and B infer Z", as well as accepting both A and B. Knowing the tortoise as we do, we know he will *still* refuse to accept Z.

It might be objected that, if after accepting the rule *R* the tortoise does not infer *Z*, that shows he does not really accept *R* as a rule. To accept a rule of inference, it might be said, is to be disposed to draw inferences in accordance with that rule. So, once the tortoise accepts *R* as a rule, he will at that point finally draw the conclusion *Z*. The moral of Lewis Carroll's story, then, is that you need more than premises to get to a conclusion. To get from the premises to the conclusion you also need a disposition or readiness to draw a conclusion from certain premises.

That seems right; but the point has no special application to logic. In particular, it gives no reason to think that the acceptance of certain logical principles involves a special readiness or disposition to draw conclusions. To show that rules of inference are to be distinguished from further premises is not to show that rules of inference have any special relation to logic. It therefore fails to show that logic is not merely a body of truths.

Similar remarks apply to what Quine says in 'Truth by Convention'.[7] Quine argues that logic cannot be said to be true by convention, because we can have only finitely many explicit conventions and there are an infinite number of logical truths. Since we cannot have a separate convention for each logical truth, we must instead formulate general conventions to cover many cases at once. But then, Quine argues, logic is needed to get logic from our general conventions. In the same way it might be argued that logic cannot be identified with a logical theory, conceived as a body of truths, since logic would be needed to get to particular logical truths from any general logical theory. But there is a serious mistake here. What is true is that *inference* is needed to get to a particular logical truth from the general logical theory. But that shows no special connection between logic and inference.

So, as far as I can see, no serious objections have ever been raised to the view that logic is a science, like physics or chemistry, a body of truths, with no special relevance to inference except for what follows from its abstractness and generality of subject matter. (I have already observed that this abstractness and generality is what accounts for the way in which logic may perhaps always be made relevant to any case in which one sees that certain of one's beliefs cannot all be true.) So it is strange that no one has unequivocably held the extreme view that logic is simply a body of truths.

That is not to say I am going to be the first to advocate that view. I feel, perhaps irrationally, that logic must have something special to do with reasoning, even if no one has yet been able to say what this might be.

In what follows, I will describe an unsuccessful attempt of mine to develop a theory which would give logic a special role in reasoning. Although this attempt is a failure, I hope it may be instructive. I suspect that any successful theory must incorporate some of the elements of my unsuccessful approach.

2. AN ATTEMPT AT A THEORY

I began with the plausible idea that, if logic has a special role to play in reasoning, that must be because it has a special role to play in the construction of arguments. For it seems that logic does play a role in argument and, furthermore, it seems that argument, calculation, and proof seem at least sometimes to facilitate reasoning. This suggested to me that, if I could understand how argument facilitated reasoning, that might begin to help me see how the use of logic might facilitate reasoning.

I did not assume at the beginning that logic must play a role in all reasoning or even in all argument. For example, I allowed for the possibility that logic is best conceived as a calculus, like the usual arithmetical calculus, or like algebra, or like the use of differential equations, so that learning logic was to learn a special technique. Learning the relevant technique would involve learning to construct arguments in a certain way. Some of one's reasoning might exploit that technique whereas other reasoning might not. Or it might be that everyone uses some sort of logic in constructing arguments, at least sometimes. I wanted to leave this question open. However logic was to be envisioned, the question worrying me was how the logical construction of arguments could facilitate belief revision.

My next thought was that implication would have to be the key. In a valid argument, the premises imply the conclusion, and each step of the argument is the acknowledgement of an implication. Furthermore, it seems that other sorts of calculation can all be treated as techniques for discovering or exhibiting implications. Given certain data, we calculate a certain result, i.e., we calculate that those data imply that result. So I thought that I might be able to understand how argument and calculation facilitate reasoning if I could understand how the appreciation of implications can facilitate reasoning. But why should the recognition of implications be important in reasoning?

This is a difficult question because in a way implication seems so *obviously* relevant to reasoning. One is inclined to wonder how anything could be *more* relevant. So here it is important to recall the points with which I began. It is not always true that one may infer anything one sees to be implied by one's beliefs. If an absurdity is implied, perhaps one should stop believing something one believes instead of accepting the absurdity. And, even if the implication is not absurd and does not conflict with other beliefs, considerations of clutter-avoidance limit how many implications of one's beliefs one can accept.

It might be suggested that implication is relevant to one's reasoning because of one's interest in believing what is true. If one believes A and B and one sees that A and B imply C, then one knows that, if one's beliefs in A and B are both true, a belief in C would also be true. So, if one has some reason to have a position on C, one has a reason to believe C. More exactly, one's reasons for believing A and B, together with one's reasons for believing they imply C, along with one's reasons for wanting to take a position on C can give one reasons to believe C.

The trouble with this suggestion is that it seems to beg the question or perhaps involve the sort of regress that afflicts the tortoise and Achilles. The suggestion seems to be that, in order to infer C from one's beliefs in A and B, one needs to believe also that, if A and B are both true, then C is also true. Then given certain other conditions one can infer that C is true, so one can safely accept C. This begs the question since it already assumes that recognition of an implication can mediate reasoning: one infers that C is true because the truth of C is implied by the truth of A and B and the proposition that A and B imply C. And a regress threatens, since the suggestion seems to imply that this last inference depends on a prior inference to the truth of "C is true," where that inference depends on another prior inference, and so on.

Furthermore, we might modify the suggestion by substituting considerations of probability for truth and turn the argument into an argument *against* inferring what is (merely) implied by what one believes. If A and B logically imply C, then the probability of C will often be smaller than either the probability of A or the probability of B. This will be true, for example, if C is the conjunction of both A and B, if A and B are even slightly independent and neither is certain. Indeed, accepting logical implications of one's beliefs could lead one to accept propositions that are quite improbable.

Probabilistic reflections of this sort might be thought to show one should never flatly believe anything but should instead assign one's conclusions varying degrees of belief. And, if one must believe things in an all-or-nothing way, it might be suggested one should adhere to a purely probabilistic rule, believing only what has a sufficiently high probability on one's evidence.

I have argued elsewhere against this sort of probabilistic approach to reasoning. Finite creatures like us cannot operate probabilistically except in very special cases. One problem is that the use of probability can involve an exponential explosion in memory and computation required. Furthermore, the practical uses to which one's basic beliefs are put constrains one's reasoning in a nonprobabilistic way to favour conclusions having to do with means to one's ends; more generally, it leads one to favor conclusions of a roughly explanatory character. One also needs to keep one's overall view of things fairly coherent. For creatures like us, inference must in a sense always be "inference to the best explanation".[8]

Now the question I have raised was how implication might be relevant to reasoning. I have been trying to show that this is a real issue whose answer is far from obvious. That is to prepare you for the hypothesis I arrived at, namely, that implication is relevant to reasoning because it is relevant to explanation.

My previous study of inference had suggested that inference is always "inference to the best explanation" or rather "inference to the best overall view", where relevant factors are conservation, coherence, and satisfaction of desires.[9] Conservatism is a factor in the sense that one should not change one's view without a positive reason for doing so and, in changing one's view, other things being equal, one ought to minimize such change. Coherence is at least in part explanatory coherence. One tries to increase the intelligibility of what one believes, trying to explain more and leave less unexplained. Because of conservatism, one tries to make minimal changes that will do this. One also tries to make minimal changes that promise to promote the satisfaction of one's desires, although that is a factor I cannot discuss here.

This is a brief sketch of the sort of conception of reasoning I had arrived at. I wanted to see how logic might be relevant to reasoning conceived in that way.

One obvious connection is that inconsistency is presumably an extreme form of incoherence. Perhaps that is why we ought to try to avoid inconsistency. It is a special case of trying to avoid incoherence, of trying to make our view as coherent as possible. The fact that coherence is not the only factor in reasoning, conservatism being another factor, would account for why it is sometimes rational to continue to believe things one knows to be jointly inconsistent. For, it may be that, in order to free oneself from all threat of inconsistency, one would have to abandon too much of one's prior beliefs, which would conflict strongly with conservatism. Conservatism can have one continue to accept inconsistent beliefs even though that leaves one with a view that is extremely incoherent in certain respects.

So, one way in which logic is relevant to reasoning may be that inconsistency is an extreme form of incoherence which reasoning seeks to avoid, other things being equal. But this does not yet address the central issue, namely, why seeing that something follows logically from one's beliefs should be relevant to accepting that consequence. It may be why one should refrain from accepting the denial of that consequence, but it does not indicate why one should sometimes positively accept the consequence itself.

At this point, I recalled that explanations often take the form of arguments. We sometimes explain something by showing that it is implied by certain other things. We sometimes understand something by seeing that it is thus implied. Reflections on this led me to the following hypothesis:

1.   We sometimes accept arguments as explanations of their conclusions.

If this is right, and if logic is specially relevant to argument, this would point to a second way in which logic was specially relevant to reasoning, over and above any relevance it may have via inconsistency.

Now, sometimes one accepts an argument as explaining something one already accepts. This is inference to the best explanation in the strict sense: one accepts something as the best explanation of one's evidence.[10] There are also cases in which one accepts an argument as explaining something one did not previously accept in terms of things one did previously accept. This is not to accept something as explaining one's evidence but is, as it were, to accept something as explained by one's evidence. Inferences about the future have this form. One infers the sugar will dissolve in one's tea because sugar is soluble in tea. One infers Albert will be in Washington tomorrow because he now intends to be there and nothing is going to change his mind or prevent him from being there.

Thinking about such cases, I was led to the following hypothesis:

2.   Whenever one infers C because C is implied by one's prior beliefs B, one accepts C as part of an argument from B to C, which one accepts as an explanation.

Hypothesis 2 is very strong. It implies that any argument that could reasonably lead one to accept its conclusion is a possible explanation.

As soon as I thought of this hypothesis, I remembered the serious objections that have been raised to deductive nomological accounts of explanation, objections which would also seem to apply to hypothesis 2. Consider Bromberger's flagpole example.[11] Bromberger observes that there is a relationship between the height of a flagpole, the angle of the sun, and the length of the flagpole's shadow. Given any two of these quantities, one can deduce the third. In one case this deduction yields an explanation, but it does not seem to do so in the other two cases. One might explain the length of the shadow by citing the height of the flagpole and the angle of the sun; but it does not seem one could normally explain the height of the flagpole by citing the length of the shadow and the angle of the sun,[12] although one could *infer* the height of the pole from the length of its shadow and the sun's angle. Nor does it seem one could normally *explain* the angle of the sun in terms of the height of the flagpole and the length of its shadow, although one could *infer* the angle of the sun from the height of the flagpole and the length of its shadow. In the last two cases, one accepts a conclusion as the conclusion of an argument that does not seem to be explanatory.

Of course, one might infer the height of the flagpole or the angle of the sun as part of the best explanation of the length of the shadow. But that would involve accepting a different explanatory argument whose conclusion was a conclusion about the length of the shadow. It would seem one can also infer the height of the flagpole more directly on the basis of an argument whose conclusion concerns the height of the flagpole, not the length of the shadow. It is this argument which does not appear to be explanatory, contrary to hypothesis 2.

Here is another example. "The man in the red shirt once climbed Mount Whitney. Jack is the man in the red shirt. So Jack once climbed

Mount Whitney." One might infer that Jack once climbed Mount Whitney on the basis of this argument. However, the argument does not explain why Jack climbed Mount Whitney, nor are there materials in the argument for explaining either why Jack is the man in the red shirt or why the man in the red shirt once climbed Mount Whitney.

In thinking about this, it occurred to me to try to distinguish causal explanations, broadly construed, which explain why something happened or why it is the way it is, from explanations that (I feel like saying) merely explain why it is true that something happened or is the way it is. I am not very happy with this way of stating the distinction, but the point is that the argument about Jack may be a kind of explanation after all. It does not explain why he once climbed Mount Whitney, what led him to do it, but it does in a sense explain why it is true that he climbed Mount Whitney. Similarly, although a calculation of the angle of the sun from the length of the shadow and the height of the pole does not explain why the sun is at that angle in the sense of explaining what causes the sun to be at that angle, the calculation does in some sense explain why it is true that the sun is at that angle.

To see that there really are noncausal explanations of this sort, consider mathematical explanations. To understand why there is no greatest prime number is to understand why it is true there is no greatest prime number. It is not to understand what causes there to be no greatest prime number. It is not to understand what leads to this being the case.

To take a different example, one of my daughters, who had been intrigued by her cousin Aaron, finally came to understand why Aaron was her cousin. "He is my cousin," she explained to me, "because his father is your brother." I believe her explanation was not of the causal sort. It explained not how it came about that Aaron was her cousin, but why it is true that Aaron is her cousin.

To take yet another example, consider explaining a particular instance in terms of a generalization. Or consider Newton's explanation of Kepler's laws. Such explanations are not causal. A generalization does not cause its instances, nor do Newton's laws show how the planet's came to observe Kepler's laws. Explanations of this sort explain why something is so without explaining what causes it to be so or what leads to its being so.

The suggestion, then, is that arguments which lead one to accept their conclusions explain why something is true rather than what caused it to happen. The contrary impression, for example, that calculating the angle of the sun does not explain why the sun is at that angle, arises because the calculation does not yield a causal explanation but only an explanation why it is true the sun is at that angle.

I repeat that this terminology is not very good. It is meant only to be suggestive. The point is that there seems to be a kind of noncausal explanation that can take the form of an argument. Arguments which lead one to accept their conclusions may be explanations of this sort rather than of the causal sort.

However, it is not important what terminology is used. The suggestion is that explanations of a sort that are clearly explanations have something in common with arguments that can figure in reasoning. They can both facilitate reasoning because they are both ways of connecting propositions that can lend coherence to one's beliefs. Although I prefer to say that arguments are explanations, which explain why something is true, anyone who objects to this terminology can interpret it as saying that arguments can induce such coherence.

But this still does not address the main issue. Even if implication is relevant to reasoning because implication can be explanatory (or can induce coherence), that is not yet to assign a special role in reasoning to logic, unless logic can play a special role in explaining why something is true (or in inducing coherence).

In thinking more about this, I considered the familiar idea that certain logical implications are immediately obvious and that other, nonobvious, logical implications can always be mediated by a series of obvious logical implications. In fully explicit arguments or proofs, each step should be an immediately obvious consequence of premises and/or previous steps. Logical rules of "natural deduction" atempt to characterize these immediately obvious steps.[13] This suggested to me that the special role of logic (if there is one) might depend on something about the immediately obvious implications that are captured by rules like the rules of natural deduction. I was therefore led to the following hypothesis:

3.　　　Certain implications can facilitate inference because they are immediately explanatory or express immediately intelligible connections of the sort that yield coherence.

All such implications are obvious. One sees immediately that they hold.

This is not to say every obvious implication is necessarily "immediate". One might be able, as it were, to combine several intermediate steps in one thought. Furthermore, one might rely on certain unstated other premises.

It may not be easy to say whether an implication is immediate in this sense. Consider, for example, "Today is Friday, so tomorrow is Saturday." Is that an immediate implication in the relevant sense, or does it involve intermediate steps and unstated premises, such as that Saturday is the day after Friday and that tomorrow is the day after today? I saw I would have to address this issue. But first I thought I should try to relate hypothesis 3 to logic. So I proposed the following hypothesis:

4.   There are basic logical implications which are immediately explanatory (immediately intelligible).

These immediately intelligible logical connections would presumably include implications covered by the basic principles of natural deduction, such as *modus ponens* ("*p* and *if p then q* implies *q*").

3. LOGIC AND LANGUAGE

I was now faced with the problem that my hypotheses were so far untestable. I had no way to distinguish immediate implications from other obvious implications; no way to say whether "Today is Friday, so tomorrow is Saturday" was an immediate implication. It was therefore unclear what it could mean to say that certain logical implications were immediate, at least if this was supposed to say more than that they were obvious implications.

Furthermore, I needed an independent criterion of logic. Otherwise it would be possible to trivialize hypothesis 4 by counting all immediate implications as logical implications. For example, if "Today is Friday, so tomorrow is Saturday" represents an immediate implication, nothing so far said would prevent it from counting as a logical implication. But, if one can in this way count every immediate implication as logical, that trivializes the claim that logic plays a special role in inference, given hypotheses 1–3.

A distinguished tradition suggests the following further hypothesis:

5.   Logical implications hold by virtue of logical form, where this is determined by the grammatical structure and logical constants involved.

So, for example, consider the following implication: "If it is raining, the picnic is cancelled. It is raining. So the picnic is cancelled." This counts as a logical implication by the present criterion since the implication holds by virtue of the logical form of the propositions involved, given that "if . . . then" represents a logical constant.[14]

Famously, Hypothesis 5 does not provide an adequate criterion of logic in the absence of a way to distinguish logical constants from other terms. Whether we should count "Today is Friday, so tomorrow is Saturday" as a logical implication depends on whether or not "today", "tomorrow", "Friday", and "Saturday" are logical constants (in a kind of tense logic). What determines the logical constants?

We could arbitrarily determine the logical constants by simply listing the terms we want to count. Somewhat less arbitrarily, we could use some sort of technical criterion to distinguish logic from other subjects and to pick out the logical constants. There are many different ways to do this, yielding many different demarcations, so that what counts as logic according to one technical criterion may fail to count as logic according to another. For example, the logical constants might be identified as those terms whose meaning can be completely determined by introduction and elimination rules of a certain sort in a Gentzen-style sequent calculus. Or they might be identified as those terms whose meaning can be completely determined by such introduction and elimination rules in a system of natural deduction. Hacking claims the former method would count the classically valid implications as logical, whereas the latter would count as logical only those that are intuitionistically valid.[15]

Now, since I was concerned with the idea that logic has something special to do with reasoning, I was disinclined to adopt an arbitrary or purely technical characterization of the logical constants. It seemed to me the identification of the logical constants should reflect something about our practices, about what our reasoning is or at least should be. So, it seemed wrong to me to appeal to the sequent calculus, which plays no obvious role in ordinary reasoning. And, although natural deduction is a plausible candidate for something that plays a role in ordinary reasoning, I was not happy to *begin* with an account of logic that immediately yielded intuitionistic rather than classical logic. And I was similarly unhappy with other relatively arbitrary or technical ways I could think of to specify the logical constants.

Now, philosophers and linguists occasionally discuss the "logical forms" of sentences of natural languages, and every once in a while it even seems that interdisciplinary collaboration in this area might prove fruitful (although this may be an illusion). This suggested to me the following hypothesis:

6.   The logical constants are grammatically distinctive in a way that has something to do with their special role in reasoning.

For example, Quine suggests that logical constants belong to small classes that are closed in the sense that it is not easy to add new terms to these classes, whereas nonlogical terms belong to large open classes to which new terms are added all the time.[16] Quine makes this suggestion about a regimented formal language, but it seemed to me the suggestion might be applicable as well to a natural language like English. The suggestion takes the logical constants to be relatively fixed, in the way that grammar is. Since the principles of logic are determined by grammatical form plus the logical constants, this suggestion would treat logic as relatively fixed as compared with theory. Changes in the logical constants and therefore changes in logic would be possible, just as changes in the grammar of the language are possible, but these changes would be unusual as compared with changes in theory due to changes in belief and in new theoretical terminology.

I liked this. It fitted in with the idea that logic has a different function from theory, so that changes in logic would involve different sorts of changes from changes in theory. It also nicely fitted in with the idea that, although one's explicit beliefs are finite, they represent infinitely many things implicitly, by implication, where the means of representation, logic, is fixed in a way that the varying explicit beliefs are not. So, I was inclined to accept Quine's suggestion and apply it to natural languages.

Stating the suggestion accurately requires some care. Not every small closed grammatical class is a class of logical constants. Prepositions in English form such a class but some of them are probably best treated as nonlogical predicates, "between" and "over" for example. What is needed is to assign words to *logical* categories, like proposition, name, one-place predicate, two-place sentential connective, quantifier, and so on. We need to consider the class of words representing atomic members of a given logical category. Then we can put forward the following hypothesis:

7.   The logical constants are those words which belong to small
         closed classes of atomic members of logical categories.

By this criterion, "and" will count as a logical constant, since the
relevant class of atomic sentential connectives is small and closed;
"between" will count as a nonlogical term, since the relevant class
of atomic relations is large and open, even if it is hard to add more
prepositions to that class. I could say more about the idea behind
hypothesis 7, but I won't since I have discussed it elsewhere.[17] For our
purposes, the details of hypothesis 7 are not important. I put the
hypothesis forward only as an example of a way to elaborate hypothesis
5.

   Notice that, however hypothesis 5 is elaborated, it will imply that
some of one's reasoning is or ought to be influenced by the language
one speaks. Some reasoning will depend in part on recognition of an
immediate logical implication, where one recognizes the implication
because it is an instance of a basic logical principle. What makes the
implication a logical implication is that it holds by virtue of logical
form, in other words by virtue of the grammatical form and logical
constants of the propositions involved. So what is relevant will be how
the propositions are expressed in language. In a sense, then, the
relevant reasoning must be reasoning in language.

   I did not mind this result. I was already inclined to believe that there
are at least two important uses of language, its use in communication
and its use in calculation or reasoning. I had earlier noticed, for
example, that, although many philosophers believe that meaning is use,
some take the relevant use to be calculation, theorizing, verification,
confirmation, and so on, and others take the relevant use to be
communication and speech acts. Surely both sides are pointing to
important aspects of language.[18]

   I should say I am not at all attracted by the idea that thought is always
in language. I believe most thought is not in language and that even
thought in language may be only partly in language and partly in some
other form of representation. Of course, by "language" here I mean a
language one speaks, like English. If we say there is a "language of
thought" (which is something I sometimes say), then the point is that, in
my view, the language of thought includes the language one speaks and
also includes other things as well.

Can there be logical constants in the language of thought that are not part of the language one speaks?[19] I saw I had better assume not. Work in linguistics gives some reason to think sentences of a natural language have a fairly determinate grammatical structure, and this work gives at least some idea of what that structure might be. There is no work of any sort that would suggest anything similar for sentences of the language of thought. So, I saw that, if I was to have any hope of avoiding unconstrained speculation, I had better assume this:

8. Logical constants are terms in a natural language one speaks.

Now, if hypotheses 5–8 are correct, it might seem we could speak of the logic of a given language, the logic determined by its grammar. But that would be true only if there was a unique grammar of the language, which furthermore determined a unique logic. This does not seem to be so. It would still seem that various competing accounts of the logic of a natural language like English are compatible with hypotheses 5–8. Consider for example how time and tense might be treated in the logic of English. One analysis might invoke a tense logic with logical constants like "past", "future", "before", "after", "today", and maybe even "Friday", and "Saturday". A different analysis would appeal to a more classical logic plus a nonlogical theory of time. It is not obvious that grammatical considerations favor one of these analyses over the other.

Another example would be how adverbs in general are to be analyzed logically. In one analysis there is an adverbial logic in which adverbial modifiers form a large open logical class. This class includes the word "not", so that "not" fails to count as a logical constant in this analysis. An alternative analysis treats most adverbs, but not "not", as predicates, so that for example the use of "suddenly" is analyzed as an application of the predicate "sudden" to an event. This dispute can be vigorously pursued, but it seems to me that grammatical considerations alone are insufficient to settle the issue.

And, as long as logic remains indeterminate, hypotheses 1–8 are themselves indeterminate and untestable.

4. COLLAPSE OF THE THEORY

Now one way to get testability and also to treat logic as having a unique role in reasoning would be to put forward the following strong hypothesis:

9.   All immediately intelligible implications expressed in language are logical implications.

This would imply that all obvious implications expressed in language either are logical implications or depend on unstated obvious assumptions, where one recognizes that the implication follows logically given those assumptions.

If hypothesis 9 is accepted together with the previous hypotheses, an empirical study of how easy it is for people to recognize implications of various sorts might help determine what the logic of our language is, since, given hypothesis 9, these empirical results might not be compatible with all the analyses permitted by grammar. The trouble is that hypothesis 9 is too strong given the earlier hypotheses. It gives us testability at the price of refutation.

The problem is that there are patterns of obvious implication that are not going to be counted as logical by the grammatical criterion and which cannot be treated as involving a series of logical steps from unstated obvious assumptions. I am thinking here of such patterns as "S knows that P, so P" or "It is true that P, so P." People do, or can come to, treat instances of these patterns as obvious implications. Furthermore, the obviousness of the implications does not depend in any significant way on the complexity of the proposition P.

This is a problem, because the implications are not just due to the grammatical forms of the propositions involved but depend on the words "know" and "true". The implications do not hold if other words replace these words. Neither "S believes that P" nor "It is unlikely that P" implies "P". But "know" and "true" are not logical constants. That is ruled out by the grammatical test. Both words are best treated as members of large open classes of atomic members of some logical category. "Know" should be classed with "believe", "hope", "fear", "expect", "regret", and so on, which are not logical constants. And "true" should be classed with "unlikely", "probable", "possible", "delightful", "surprizing", and so on.

So our hypotheses imply that these patterns of logical implication are not logical patterns. That means, by hypothesis 9, they depend on hidden premises, on unstated obvious assumptions. The needed premises will be part of a theory of knowledge or of truth.

These premises either are explicitly believed or are themselves obvious consequences of other things explicitly believed, in which case the latter things are the basic assumptions appealed to. Since one's explicit beliefs must be finite, recognition of instances of these patterns of implication must depend on acceptance of some sort of finite theory of knowledge and theory of truth. Now, it would seem that, given anything like classical first-order quantificational logic, a finite theory of either sort adequate to account for all instances of the relevant pattern will require connecting arguments that become more complex as P becomes more complex. So, given first-order logic, hypothesis 9 predicts a fairly rapid decline in obviousness for instances of the pattern as the complexity of P increases. As already noted, that prediction is false. Therefore, our hypotheses require that the logic of the language, in this case English, cannot be classical first-order quantificational logic. We need to assume a logic which will allow each instance of the pattern to depend fairly simply on some unstated assumption.

More precisely, we need a second-order logic that would allow one to express the assumption, "For all P, if someone knows that P, then P." With that as an implicit assumption, each implication of the pattern can be derived in two steps, first getting the relevant instance of the implicit assumption, then applying *modus ponens*.

The trouble is that English does not appear to involve second-order logic, at least in this way. The crucial assumption, "For all P, if someone knows that P, then P," is not as stated expressed in ordinary English. Indeed it would seem to have no ordinary means of expression. The closest one can come is "If someone knows something, it is so." But that would seem to be a purely verbal variant of "If someone knows something, it is true." That is not what we want, as is evident from the fact that we need also to account for implications of the pattern "It is true that P, so P."

It is sometimes argued that English *does* allow higher-order quantification. If that were so, it would be some confirmation of the hypotheses I have stated. But I am doubtful. Strawson notes that we say things like "Albert is everything one would want in an assistant."[20] However, that construction is marginal and its interpretation is obscure. Grover, Camp, and Belnap suggest that "it is true" might function as a "pro-sentence", which is related to a sentence as a pronoun is to a noun phrase, functioning logically as a variable taking sentence position.[21] This would imply that the statement, "If someone knows something, it is true," is the way to express in English the proposition. "For all P, if someone knows that P, then P." But this seems wrong as an account of how "true" functions in English.[22]

So, I reluctantly conclude that some immediate implications are nonlogical, the ones just discussed and perhaps many others as well. This means that hypothesis 9 must be rejected; logic is not the only source of immediate implications. But then we cannot constrain logico-grammatical analysis by means of data concerning the implications people find obvious. So, it remains unclear whether the grammatical criterion will single out a unique logic of English. And that means we will not be able to say what the boundaries of logic are. So we end up with no nonempty hypothesis concerning the special role of logic in reasoning. This attempt to make sense of that idea ends in failure. Whether there is some other way to make sense of that idea I cannot say.

NOTES

[1] "The representation of logic as conceived with a characteristic of sentences, truth, rather than of transitions from sentences to sentences, had deleterious effects both in logic and philosophy." Michael Dummett, *Frege: Philosophy of Language*, Duckworth, London, 1973, pp. 432–433. Dummett's remark is quoted with approval by Ian Hacking, for whom "logic is the science of deduction" in accordance with "rules of inference" like Gentzen's introduction and elimination rules. See Ian Hacking, 'What Is Logic?' *Journal of Philosophy* **76**, 285–319, quoting from pp. 292–293; and Gerhard Gentzen, 'Investigations into Logical Deduction', a translation of a 1935 paper by M. E. Szabo in *The Collected Papers of Gerhard Gentzen*, North Holland, Amsterdam, 1969, pp. 68–131. See also W. C. Kneale, 'The Province of Logic', in H. D. Lewis, *Contemporary British Philosophy*, Third Series, Allen and Unwin, London, 1956.

[2] For an example, see my "Reasoning and Evidence One Does Not Possess', *Midwest Studies in Philosophy* **5**, 1980, 163–182, esp. p. 172.

[3] Gottlob Frege, *Grundgesetze* (1893), partial translation by M. Furth, *The Basic Laws of Arithmetic*, University of California Press, Berkeley, 1967, pp. 12–13.

[4] W. V. Quine, *Methods of Logic*, Holt, Rinehart, and Winston, New York, 1972, pp. 1–5.

[5] Quine, *Methods of Logic*, p. 39, and 'Truth by Convention', in *The Ways of Paradox*, Random House, New York, 1966, pp. 70–99. I will say more about this article in the text below.

[6] Lewis Carroll, 'What the Tortoise Said to Achilles', *Mind* **4**, 1895, 278–280.

[7] Cited in note 5.

[8] Gilbert Harman, 'Reasoning and Explanatory Coherence', *American Philosophical Quarterly* **17**, 1980, 151–159.

[9] Actually, it is probably an exaggeration to say "best" here. Herbert Simon observes

that it is often more reasonable to "satisfice" than to maximize. See Herbert A. Simon, *Administrative Behavior*, 3d ed., The Free Press, New York, 1976, e.g., p. xxviii; 'A Behavioral Model of Rational Choice' reprinted in Herbert A. Simon, *Models of Thought*, Yale Univ. Press, New Haven, 1979, e.g., p. 11. It may be better to say one infers a satisfactory explanation, if one can think of one, without one's normally attempting to find the best explanation.

[10] Perhaps it would be more accurate to say this is inference to *satisfactory* explanation of one's evidence.

Here, of course, I am using the term "explanation" to refer to what one grasps when one understands why something is, was, or will be the case. I say one grasps an explanation when one grasps something of the form "*A* because *B*." There is another sense in which an explanation is a speech act, an *explaining*. In the second sense, one can explain only what one already believes to be the case. But in the first sense, in which I am using the term "explanation," one can come to see that something one already knows about will be responsible for and will therefore account for and explain something about which one was previously ignorant.

[11] Sylvain Bromberger, 'Why Questions', in R. G. Colodny (ed.), *Mind and Cosmos*, Univ. of Pittsburgh Press, Pittsburgh, 1966.

[12] Bas van Fraassen observes that such an explanation might be possible if the pole was constructed so as to cause a shadow of that length when the sun is at that angle. But, of course, that is not the usual case. See Bas van Fraassen, 'The Pragmatics of Explanation', *American Philosophical Quarterly* 14, 1977, 143–50.

[13] See Gentzen, 'Investigations into Logical Deduction'.

[14] And given that "if ... then" represents a sentential connective. In fact, there are reasons for a different analysis in which *modus ponens* is not a principle of logic. See Gilbert Harman, "If and Modus Ponens: A Study of the Relations between Grammar and Logical Form', *Theory and Decision* 11, 1979, 41–53.

[15] Ian Hacking, 'What is Logic?', 292–293.

[16] W. V. Quine, *Philosophy of Logic*, Prentice-Hall, Englewood Cliffs, 1970, pp. 28–29, 59.

[17] Gilbert Harman, 'If and Modus Ponens'.

[18] Gilbert Harman, 'Three Levels of Meaning', *Journal of Philosophy* 65, 1968, 590–602.

[19] Here I am indebted to Kit Fine.

[20] P. F. Strawson, *Subject and Predicate in Logic and Grammar*, Methuen, London, 1974, p. 33.

[21] D. L. Grover, J. C. Camp, and N. D. Belnap, 'A Prosentential Theory of Truth', *Philosophical Studies* 35, 1979, 289–297.

[22] Sarah Stebbins, 'Necessity and Natural Languages', *Philosophical Studies* 35, 1979, 289–297.

Dept. of Philosophy
Princeton University
Princeton, New Jersey 08544
U.S.A.

JOHN MYHILL

PARADOXES

1. "PARADOXES" OF SET THEORY AND PARADOXES OF PROPERTY THEORY

Gödel said to me more than once, "There never were any set-theoretic paradoxes, but the *property-theoretic* paradoxes are still unresolved"; and he may well have said the same thing in print. This paper is concerned with the *second* problem, the property-theoretic paradoxes which arise from Frege's conceptions; but before I begin to deal with these, a few comments about the first half of Gödel's remark are in order. This is because there is at present a serious misunderstanding, so prevalent that it is almost universal, about the Cantorian notion of set and its allegedly paradoxical properties; and I cannot pass by this misunderstanding unnoticed.

The current heresy bespeaks at once a (hopefully temporary) victory of pragmatism and generally sloppy thinking over philosophical analysis, and at the same time a failure of literary and exegetic scholarship in the traditional sense. It holds that Cantor had a "naïve" conception of set according to which every well-formed formula with one free variable determines a set; and that the paradoxical consequences of this conception were swept under the rug by Cantor himself and repaired afterwards by various ad hoc principles designed to yield all the set theory needed for mathematics and still hopefully preserve consistency. This flies directly in the face of the evidence of the text; Cantor, who discovered most of the so-called paradoxes himself, never regarded them as a source of trouble, but in fact *used* one of them (the Burali-Forti "paradox", which he discovered before Burali-Forti) as a *lemma*, perfectly correctly, in his proof of the well-ordering theorem (which he discovered before Zermelo). The proof runs as follows: Suppose S is a set which cannot be well-ordered. Define a one-one map f from the ordinals into S as follows: $f(0)$ is to be any element of S (which is certainly nonempty), and if f has been defined for all ordinals less than some ordinal α, $f(\alpha)$ is some element of S which has not yet been assigned as an image to any such ordinal. There must be such an

Synthese **60** (1984) 129–143.   0039–7857/84/0601–0129 $01.50
© 1984 *by D. Reidel Publishing Company*

element, since if all of S were used up, S would be well-ordered *contra hypothesin*. Hence the range S' of f would be in one-one correspondence with the ordinals; since by the Burali-Forti paradox the ordinals are not a set, neither (by the axiom of replacement) is S'; a fortiori S, which includes S', cannot be a set, contradiction. The details of this proof are to be found in Cantor's famous letter to Dedekind, which is reprinted at the end of Zermelo's edition of Cantor's collected works; and the required properties of sets, including the existence of "inconsistent totalities", are made quite clear in mathematical and philosophical works of Cantor which antedate his letter to Dedekind by more than a decade. This is not the place to discuss what Cantor's conception of a set *was*, but the texts themselves make it quite clear that it was not naïve but, on the contrary, quite sophisticated, and never for a moment confused in Cantor's mind with the Fregean concept of "property" (Cantor's "Vielheit" or "Inbegriff"). I repeat the crucial historical points: (a) the distinction of Vielheit or Inbegriff vs. Menge long antedates the discovery of paradoxes and (b) when (the Burali-Forti) paradox appeared, it was used as a *lemma* by Cantor for a theorem he wanted to prove, not regarded as a disaster or even a surprise. Professor Joseph Dauben, the only person to write a full-length book on Cantor's work, claims otherwise, and is surprised that Cantor "regarded the paradoxes as beneficial" (p. 243). Since it is abundantly clear from the text that for Cantor they were *not* paradoxes, one can only conclude that Professor Dauben's intelligence and sympathy in reading those texts are of a piece with his mathematical acumen when he states on another page (p. 187) that the axiom of choice is needed to well-order a *countable* set.

2. FITCH'S APPROACH: DESCRIPTION OF K'

Enough of polemic. Let us turn to the main concern of this paper, the *property-theoretic* paradoxes which according to Gödel are still with us, and which follow from what I call "Frege's principle", that every formula with one free variable determines a *property* (not a set) which holds of all those and only those things which satisfy the formula. If the formula in question is "$x \notin x$" (where as henceforth we read "$\in$" not as "is a member of" but as "has as a property") we get the Russell paradox, and from this it follows that the Fregean concept of property is inconsistent *with classical logic*. So if we want to take Frege's principle seriously, we must begin to look at some kind of nonclassical logic. The first serious attempt to do this was made by my old colleague F. B. Fitch in "An Extension of Basic Logic'. He proceeds by defining a class of expressions called U-*expressions*, and then singles out from these, by a simultaneous inductive definition, two subclasses called the *true* and *false sentences*. The procedure is simplicity itself: variables are terms, and if s and t are terms then $s = t$ and $s \in t$ are formulas. Terms are built from formulas by the abstraction symbol $\{x \mid \ldots\}$. Other formulas are built up in the standard way by connectives $\wedge$ and $\neg$ and quantifiers $(\forall x)$. Existential quantification and disjunction receive their classical definitions. The inductive definition of truth and falsehood for sentences (formulas without free variables) is completely straightforward: $s = t$ (closed) is true iff s and t are literally the same expression, and false otherwise (so that extensionality fails very badly; there are other possibilities); $A \wedge B$ is true iff both A and B are true, and false if at least one of them is false; $\neg A$ is true or false according as A if false or true; $(\forall x)(A(x))$ is true iff all its closed instances are true, and false iff at least one such instance is false. Finally, $t \in \{x \mid A(x)\}$ receives always the same truth-value (if any) as $A(t)$. It is immediate that no sentence receives values of *both* T and F under this definition, and that many (e.g., $\{x \mid \neg(x \in x)\} \in \{x \mid \neg(x \in x)\}$) do not receive either.

I have for the sake of clarity presented these matters in a notation different from Fitch's though essentially equivalent to it. I have also omitted the notions of formula, truth, and falsehood in regard to *arithmetical* statements; I shall return to this later since it is important. But for the moment I present only the strictly logical and property-theoretic part of the system, because I want you to realize the utter simplicity of it, and of the clarity of the interpretation (far clearer and more unambiguous than that of set theory) upon which it is based.

I said "system" and this was a slip; so far we have given only the *interpretation* or truth-definition. The next order of business is to formalize (a part of) K' as a deductive *system*, and to this we now turn.

3. NATURAL-DEDUCTION SYSTEM FOR K'

I presume here some familiarity with natural-deduction systems; a proof consists of a sequence not only of *lines* (formulas or sentences) but also of *subproofs* some of which may contain lines designated as *hypothesis*. Every line of a proof or subproof is either (a) a *hypothesis*, (b) an *axiom*,

(c) a line of a subproof *S reiterated* into that subproof from a proof of which *S* is an item, or (d) a consequence of preceding items (lines or subproofs) of the given proof or subproof by one of the *rules of inference*.

The axioms and rules of inference of the subsystem, call it K_0, of Fitch's K', which we are presenting, are the following.

1. IDENTITY
Axioms

$$s = s$$

$$s = t \lor \neg(s = t)$$

Here and henceforth *s*, *t*, etc., will denote (open and closed) *terms*. $A \lor B$ receives its classical definition $\neg(\neg A \land \neg B)$

Rule

$$\frac{a = b, \ A(a)}{A(b)}$$

2. PROPOSITIONAL CALCULUS
Rules

$$\frac{A, B}{A \land B} \qquad \frac{A \land B}{A} \qquad \frac{A \land B}{B}$$

$$\frac{\neg A}{\neg(A \land B)} \qquad \frac{\neg B}{\neg(A \land B)}$$

$$\frac{\neg(A \land B) \quad \begin{array}{|l} \underline{\neg A} \\ C \end{array} \quad \begin{array}{|l} \underline{\neg B} \\ C \end{array}}{C}$$

This last rule says that *C* is a consequence of three things: the formula $\neg(A \land B)$ and two subproofs, one having $\neg A$ as a hypothesis and *C* as an item, the other having $\neg B$ as a hypothesis and *C* as an item. The little lines under $\neg A$ and $\neg B$ indicate the status of these formulas as hypotheses. The purpose of this rule is to allow (via the above definition of disjunction) the use of the derived rule:

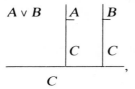

traditionally known as "Dilemma"

$$\frac{A}{\neg\neg A} \qquad \frac{\neg\neg A}{A} \qquad \frac{A, \neg A}{B}$$

3. QUANTIFICATION

$$x \begin{array}{|c} \vdots \\ A(x) \end{array} \qquad \frac{(\forall x)A(x)}{A(t)}$$
$$\frac{}{(\forall x)A(x)}$$

The "x" to the left of the subordinate proof of $A(a)$ in the first rule indicates that that proof is *general* with respect to x; i.e., that it remains a proof when any closed term is substituted for x. As is well-known, it is possible to formulate this condition in purely syntactical terms.

$$\frac{\neg A(t)}{\neg(\forall x)A(x)} \qquad \neg(\forall x)A(x) \qquad x \begin{array}{|c} \neg A(x) \\ \\ B \end{array}$$
$$\frac{}{B}$$

The subordinate proof of B in the last rule is general with respect to x, and it has $\neg A(x)$ as hypothesis. The rule looks more familiar in the form

$$(\exists x)A(x) \qquad x \begin{array}{|c} A(x) \\ \\ B \end{array}$$
$$\frac{}{B} ,$$

(existential-quantifier elimination) which follows from it if we use (as we shall henceforth) the classical definition of $(\exists x)$ as $\neg(\forall x)\neg$.

$$\frac{(\forall x)(A(x) \vee B)}{((\forall x)A(x)) \vee B}$$

The main purpose of this rule is to derive the (probably weaker) rule

$$\frac{(\forall x)(A(x) \vee \neg A(x))}{(\forall x)A(x) \vee \neg(\forall x)A(x)}$$

which, together with the other rules of K_0, allows us to use classical logic on formulas built up out of atomic formulas $s = t$ by connectives and quantifiers. This will become important when we come to discuss arithmetic.

4. ABSTRACTS

$$\frac{A(s)}{s \in \{x \mid A(x)\}} \qquad\qquad \frac{\neg A(s)}{(s \in \{x \mid A(x)\})}$$

The double inference-lines (══════) in these rules signify that the inference is permissible in either direction.

It is straightforward to show that every sentence provable in K_0 is true in K' in the sense of the truth-definition. The converse is also easily seen to be true if we adjoin to K' the (infinitary) *ω-rule*

$$\frac{A(t_1), \ A(t_2), \ A(t_3), \ldots}{(\forall x)A(x)}$$

when $t_1, t_2, t_3, \ldots$, are all the closed terms there are.

In order to discuss the strengths and weaknesses of K_0 and K', we shall want to adjoin to both *natural numbers* and the usual elementary operations (e.g., $+$ and $\times$) on them. Formally we add the symbols N, 0, s (successor), $+$, $\times\ldots$ and extend the truth-definition for K' by requiring that $t \in N$ is true or false according as t is or is not a (closed) numerical expression, and that $s = t$ is true iff s and t can be converted into one another by evaluation of closed numerical expressions (so that e.g., $\{x \mid 3 + 1 \in x\} = \{x \mid 2 \times 2 \in x\}$ is true) – and false otherwise. In K_0 we add the recursive definitions of $+$, $\times$ and whatever other functions we may want, together with Peano's axioms in the form

$$0 \in N \qquad t \in N \vee \neg(t \in N) \qquad \frac{t \in N}{s(t) \in N} \qquad \neg(0 = s(t)) \qquad \frac{s(s) = s(t)}{s = t}$$

and

$$A(0) \qquad x \ \big\lfloor A(x) \wedge x \in N \qquad t \in N$$

$$A(\mathrm{s}(x))$$

$$\overline{\qquad\qquad\qquad\qquad\qquad\qquad}$$

$$A(t)$$

It is then straightforward to show that the augmented K_0 (called $K_0 + \text{arith}$) is contained in the augmented $K' + \text{arith}$ in the same sense in which the original K_0 was contained in K'. In particular $K_0 + \text{arith}$ *is consistent*.

4. MATHEMATICS IN $K_0 + \text{arith}$

Not only is it consistent, but it contains a fair amount of mathematics. This has been developed by Fitch in a series of articles and something similar was done by Schütte in his *Beweistheorie*. In the first place, since as pointed out above we have classical logic for formulas built out of identities by connectives and quantifiers, and since we have Peano's axioms, the whole of Peano arithmetic is forthcoming in $K_0 + \text{arith}$, and by the same token *all* arithmetical truths in $K' + \text{arith}$. What about analysis?

Fitch showed that if we define real numbers as Dedekind cuts (to define them as Cauchy sequences yields an (interesting) non-classical theory) we are in serious trouble. Specifically, if we let R be the property of being a real number so defined, we do *not* have $(\forall x)(x \in \mathbf{R} \vee x \in \mathbf{R})$. This complicates the development of analysis for the following reason. Many of the simplest formulas of analysis (or rather high-school algebra), e.g., $(\forall x)(\forall y)(x \in \mathbf{R} \wedge y \in \mathbf{R} \supset x \le y \vee x > y)$ become unprovable and even lead to contradiction, at least if we define $A \supset B$ as $\neg A \vee B$. This is not surprising, as we cannot even prove (or add consistently as an axiom) $(\forall x)(x \in \mathbf{R} \supset x \in \mathbf{R})$ on *this* definition of implication. The best we can do apparently is to prove as *metatheorems* such statements as: if $t_1 \in \mathbf{R}$ and $t_2 \in \mathbf{R}$ are theorems, then either $t_1 \le t_2$ or $t_1 > t_2$ is a theorem. This development of analysis in the metalanguage was carried out by Fitch, in the series of papers just mentioned, as far as the Heine-Borel theorem; the elementary properties of the derivative and (Riemann) integral present no problem. However there is a great inelegance in this treatment, even if (as Schütte did) we formalize the metalanguage too. After all, we would like *the system itself* to contain analysis, we would like it to be able to state and prove that every real number is positive, negative or zero, and so on. For this, evidently, we need some sort of implication $((\forall x)(x \in \mathbf{R} \supset \cdots))$; and all the above-mentioned bad results show is that it can't be the classical $\neg A \vee B$.

5. IMPLICATION – THE PROBLEM

Let us consider how to introduce implication into the system. The most essential properties of implication, accepted by the classical logicians and intuitionists alike, are *modus ponens* and the *deduction theorem*. Formally these are

$$\frac{A, \, A \supset B}{B}$$

and

$$\frac{\begin{array}{|l} A \\ \\ B \end{array}}{A \supset B,}$$

respectively.

There are two objections to the obvious procedure of adding a new sign $\supset$, with these two rules, to, e.g., the system $K_0 + \text{arith}$; and of these the *least* important is that the resulting system is inconsistent. First let us satisfy ourselves that this is indeed so.

Let $\perp$ be some sentence refutable in $K_0 + \text{arith}$, e.g., $1 = 0$, and let C (for Curry) abbreviate $\{x \mid x \in x \supset \perp\}$. We have as a proof in the strengthened system

$$
\begin{array}{ll}
\left. \begin{array}{l} C \in C, \text{ i.e., } C \in \{x \mid x \in x \supset \perp\} \\ \quad C \in C \supset \perp \\ \quad \perp \end{array} \right. & \text{(hypothesis)} \\
\quad\quad\quad\quad\quad\quad\quad\quad & \text{(abstract rule)} \\
\quad\quad\quad\quad\quad\quad\quad\quad & \text{(modus ponens)} \\
C \in C \supset \perp & \text{(deduction theorem)} \\
C \in \{x \mid x \in x \supset \perp\}, \text{ i.e., } C \in C & \text{(abstract rule)} \\
\perp & \text{(modus ponens)}
\end{array}
$$

Parenthetically this shows that though some form of nonclassical logic is necessary to avoid the property-theoretic version of Russell's paradox, intuitionistic logic (which includes both our implication rules) is not the right track to take.

But this inconsistency is only a symptom of a deeper malaise. What exactly have we done? We have adjoined a couple of rules to K_0 + arith, and deduced a contradiction. In doing this we have been guilty of formalism, of behaving in the way in which according to the current heresy Zermelo and Fraenkel behaved, i.e., of putting down axioms pragmatically without regard to an interpretation. Let us then backtrack a little, and try to explain what implication *means* before trying to axiomatize it.

Explaining what things *mean* is done conventionally by exhibiting a *model*; but since we are here in the domain of nonclassical logic for which a notion of model is not at hand, we should proceed by the alternative method (substantially equivalent in the classical case) of giving a *truth-definition*. This is in fact what we have been doing all along so far; we explained first the truth (and falsity) of atomic formulas $s = t$ and $N(s)$, and then we explained the truth (and falsity) of $A \land B$ in terms of truth and falsity of A and of B, the truth and falsity of formulas $(\forall x)A(x)$ in terms of the truth and falsity of their instances $A(t)$ and so on. It is our task now to extend this truth definition to formulas $A \supset B$; let us recall why. The definition $A \supset B \equiv \neg A \lor B$ does not give us the kind of implication we need in mathematics, because we don't have $A \supset A$ for A's which are neither true nor false, and so we are unable to restrict quantifiers to conditions like '$x \in \mathbf{R}$' which are not true or false for all x.

It might seem at first that giving a truth-definition for implication is a hopeless task, for any possible candidate must, by the above argument, fail to satisfy either modus ponens or the deduction theorem, and one might wonder why a relation, however defined, which lacked either of these crucial properties, should deserve the name of implication at all.

6. THE "ENTERPRISE" SOLUTION

My solution consists of two observations: firstly, that in the absence of classical logic it is reasonable to identify implication with *deducibility* and secondly, that we must specify *what the rules are* which are referred to in the notion of deducibility. Let us now look at the above paradox about $\{x \mid x \in x \supset \bot\}$ in this light. We have

1. $\quad \lfloor \{x \mid x \in x \supset \bot\} \in \{x \mid x \in x \supset \bot\}$
2. $\quad \mid \{x \mid x \in x \supset \bot\} \in \{x \mid x \in x \supset \bot\} \supset \bot$
3. $\quad \mid \bot$
4. $\{x \mid x \in x \supset \bot\} \in \{x \mid x \in x \supset \bot\} \supset \bot$
5. $\{x \mid x \in x \supset \bot\} \in \{x \mid x \in x \supset \bot\}$
6. $\bot$

Line 2 is obtained from line 1 by abstract elimination; line 3 from lines 1–2 by modus ponens, and line 4 from the subproof 1–3 by implication introduction (deduction theorem). Already the essence of the trouble is revealed. If implication means deducibility by certain rules (i.e., if one of the clauses of our truth-definition is '$A \supset B$ is true iff B can be obtained from A by abstraction introduction, universal-quantifier elimination, etc.') we cannot without circularity include amongst those rules *the rules for that very implication which we are trying to define*. There are perfectly good notions of deducibility to hand already, for example, deducibility in K_0 + arith together with the ω-rule; this notion satisfies modus ponens, but it only satisfies the deduction theorem if we use only the rules of K_0 + arith and the ω-rule in the deduction, in particular if we do not use any implication rules therein.

Let us denote this by $\supset_0$, and let C be short for $\{x \mid x \in x \supset_0 \bot\}$. Assume $C \in C$, then $C \in C \supset_0 \bot$ by abstract elimination, and $\bot$ follows from the two of them by *modus ponens for* $\supset_0$ which is *not* one of the rules deducibility by which $\supset_0$ is meant to formalize. $C \in C$ does indeed imply $\bot$ but in a weaker sense of implication which we shall call $\supset_1$. In general $\supset_0$ means deducibility without the use of any implication rules at all, and $\supset_{n+1}$ means deducibility using no implication rules except those governing $\supset_0, \supset_1, \ldots, \supset_n$.

I attempted to formalize this notion many years ago in a short note in *The Logical Enterprise*, a Fitch memorial volume. Here is the formalization (which is meant to be added to essentially K_0 + arith).

Rule of implication introduction ($\supset_n$ *int*).

$$\frac{\lfloor A \\ \quad \mid B}{A \supset_n B,}$$

where in the deduction of B from A (including the proofs of any items reiterated into it) no use is made of any of the rules $\supset_m$ int, $\supset_m$ elim, neg $\supset_m$ int or neg $\supset_m$ elim for $m \geq n$.

Rule of implication elimination ($\supset_n$ elim).

$$\frac{A,\ A \supset_n B}{B}$$

Rule of negative implication elimination ($\neg \supset_n$ elim).

$$\frac{A,\ \neg B}{\neg(A \supset_n B)}$$

Rule of negative implication elimination ($\neg \supset_n$ elim).

$$\text{(a)} \quad \frac{\neg(A \supset_n B)}{A} \qquad \text{(b)} \quad \frac{\neg(A \supset_n B)}{\neg B}$$

It is evident that this information blocks the above "proof" of $\bot$ effectively; if we write $\supset_0$ in lines 1–2, line 4 must give way to

$$\{x \mid x \in x \supset_0 \bot\} \in \{x \mid x \in x \supset_0 \bot\} \supset_1 \bot$$

and line 5 to

$$\{x \mid x \in x \supset_0 \bot\} \in \{x \mid x \in x \supset_1 \bot\},$$

from which there is no apparent way to deduce anything unpleasant. It also appears that the given formalism does justice to the informal motivation above for distinguishing "levels" of implication. However, in the "Enterprise" paper I was not able to give a consistency proof, let alone an interpretation (truth-definition) for this system, and things bogged down for several years.

7. THE CONSISTENCY PROOF

I and my student Bob Flagg (who proved a crucial and difficult Lemma) were able to take care of this situation: what we did was to extend $K' +$ arith by a Gödel numbering of formulas and then *define* in the extended system implications $\supset_0, \supset_1, \supset_2, \ldots$ satisfying the above rules. The rules for Gödel numbering are the obvious ones: if t is a term and A is a closed formula (sentence), then $G(t, A)$ is a formula; $g(t, A)$ is an axiom if t is a closed term denoting the Gödel number of the sentence A, and $\neg G(t, A)$ is an axiom for any other closed t and A. Call this system $K' + G$; it is defined purely by introduction rules, in other words by a truth (and falsity) definition.

Say that a set α of natural numbers is *represented* in K' (or $K' + G$) if $n \in \alpha$ is equivalent to K' (or $K' + G) \vdash N \in t$ for some closed term t. Lorenzen and I have shown that the representable sets of natural numbers are precisely the Π_1^1 sets: the terminology and result extend trivially to relations. In particular, the relation between the Gödel numbers of the premiss and the conclusion of a proof in $K_0 + G$ (or better $K_0 + G +$ the ω-rule) is represented by a formula $\mathrm{imp}_0(m, n)$ of $K' + G$; so we define

$$A \to_0 B \equiv (\exists mn)(G(m, A) \wedge G(n, B) \wedge \mathrm{imp}_0(m, n)).$$

For this kind of implication we have immediately modus ponens and the deduction theorem, provided in the latter case that no implication rule is used in the deduction, i.e., we have the analogs of $\supset_0$ int and $\supset_0$ elim for $\to_0$. By Flagg's complex and technical argument, which we shall give elsewhere, it is possible to modify the definition of $\to_0$ to obtain a connective $\supset_0$ which satisfies the two negative rules also.

Now we extend $K_0 + G$ (with or without the ω-rule) to $K_0 + G$ ($+$ the ω-rule) $+$ the four $\supset_0$ rules; the deducibility relation in this system is again Π_1^1, so we represent it in $K' + G$ by a formula $\mathrm{imp}_1(m, n)$ in terms of which $\to_1$ and $\supset_1$ are defined exactly as $\to_0$ and $\supset_0$ were defined from imp_0. Iterating this procedure a countable number of times, we get a system containing $\supset_n$ and $\neg \supset_n$ int and elim for all n. Notice that $\supset_n$ has a perfectly definite meaning and is not merely an ad hoc device to avoid contradiction; the meaning of $A \supset_n B$ is given by the two clauses (1) it is true iff B is deducible from A using no implication rules but those for $\supset_0, \ldots, \supset_{n-1}$ and (2) it is false iff A is true and B is false.

Let us recapitulate the contribution in somewhat different and more general terms. One starts with a language L (in our case $K' + G$) containing only introduction rules (or equivalently a truth-definition) and with no implication at all. One then conservatively extends it to L_0 with elimination rules too; "conservatively" means if $L_0 \vdash A$ then $L \vdash A$, and if $A \vdash B$ in L_0 and $L \vdash A$ then $L \vdash B$. In other words, one constructs L_0 so as to ensure certain closure properties of L. One then defines in L the deducibility relation of L_0 and calls it $\supset_0$. Then one extends L_0 to L_1 by adjoining further introduction and elimination rules for $\supset_0$ and defines the deducibility relation of L_1 in L, calling it $\supset_1$. And so on and so on. The exact details of construction of L_0, L_1, ... do not matter, provided that firstly, they are conservative over L in the sense described, and secondly, their deducibility relations are definable in L.

8. SOME ADDITIONS

This leaves some leeway in the construction of the system. So far, we have added to $L = K' + G$ only the obvious rules for $\supset_n$, positive and negative introduction and elimination. However, there are some obvious further closure properties which we have not codified. For example, if $A \supset_0 B$ and $B \supset_0 C$ are provable in L_1, so is $A \supset_0 C$; we can therefore add to L_1 the rule

$$\frac{A \supset_0 B, \ B \supset_0 C}{A \supset_0 C} \ .$$

This does not increase the stock of theorems of L_1, but it changes its deducibility relation and therefore the meaning of $\supset_1$. So we now have as a theorem

$$((A \supset_0 B) \wedge (B \supset_0 C)) \supset_1 (A \supset_0 C),$$

and in the same way

$$((A \supset_n B) \wedge (B \supset_n C)) \supset_{n+1} (A \supset_n C),$$

which we did not have before. Likewise we can add

$$(\forall x)(A \supset_n F(x)) \supset_{n+1} (A \supset_n (\forall x)F(x)),$$

and presumably many other things too. Formally we adjoin the rules

$$\frac{A \supset_n B, \ B \supset_n C}{A \supset_n C}$$

and

$$\frac{(\forall x)(A \supset_n F(x))}{A \supset_n (\forall x)F(x)},$$

and we do not permit the use of these rules in subordinate proofs used for $\supset_k$ int with $k \leq n$.

The amount of leeway about what additional rules we can add may certainly be felt as a disadvantage. The truth is of course that there are very many different notions of implication depending on what rules of inference we have, and that the rules of inference are not determined by the set of theorems, so that the systems with or without the above rules, or with one of them but not the other, are all equally good, i.e., sound and complete for the appropriate notion of implication. However, it *would* be nice to have some objective reason to prefer one of those systems over the rest. This could be done if we had a set of implication rules which in some sense was complete; but it is not yet clarified how to define such a notion of completeness.

9. CONCLUSION

Now let us see what the resulting system looks like as a vehicle for mathematics. If we use $(\forall x)(x \in \mathbf{R} \supset_0 \ldots)$ for quantifiers restricted to real numbers and $(\forall f)((\forall x)(x \in \mathbf{R} \supset_0 f(x) \in \mathbf{R}) \supset_1 \ldots)$ for quantifiers restricted to real functions we get a body of analysis considerably more extensive than that contained in Bishop's book (apart from measure theory, which presumably must be developed along the lines of Bishop and Cheng). In particular, we can prove the Heine-Borel theorem, König's lemma and Peano's theorem on differential equations.

It seems possible that the self-referential features of the system will permit a decent unstratified form of category theory (without the distinction of "small' and "large" categories) to be worked out, but there is much work to be done here. We have in any case achieved a system with a clear semantics rather than ad hoc devices to avoid contradictions, based on Frege's principle and adequate to a substantial body of mathematics.

10. POSTSCRIPT ON SEMANTICAL PARADOX

The presence of the "Gödel number relation" G which we introduced merely as a technical device to define implication also makes it possible to dispel the paradoxes typified by the "liar" and heterological paradoxes. For example, the set of Gödel numbers of true sentences is easily seen to be Π_1^1 and there is consequently a formula $T(x)$ which represents it. By the standard Gödel trick we can produce a sentence E (the "Epimenides") such that

$$\vdash E \equiv_0 \neg T(\ulcorner E \urcorner),$$

(where $A \equiv_0 B$ is $(A \supset_0 B) \land (B \supset_0 A)$); likewise we can define a term Het (heterological) such that for each closed term t

$$\vdash t \in \mathrm{Het} \equiv_0 (\exists x)(G'(t, x) \land t \notin x),$$

(where G' represents the relation of the Gödel number of a closed term to the term itself), so that in particular

$$\vdash \ulcorner \mathrm{Het} \urcorner \in \mathrm{Het} \equiv_0 \ulcorner \mathrm{Het} \urcorner \notin \mathrm{Het}.$$

There are no paradoxes since sentences like E and $\mathrm{Het} \in \mathrm{Het}$ are neither true nor false. The resulting "internal" semantics is substantially the same as Kripke's theory of truth; but it's better because we can actually state *in the system* that, e.g., the Het-paradox is a paradox of order 0. (A paradox of order n is a sentence A for which $\vdash A \equiv_n \neg A$.)

REFERENCES

Anderson, A. R. and R. B. Marcus: 1974, *The Logical Enterprise*, Yale Univ. Press, New Haven.
Bishop, E.: 1967, *Foundations of Constructive Analysis*, McGraw-Hill, New York.
Bishop, E. and H. Cheng: 1972, *Constructive Measure Theory*, American Mathematical Society Memoirs 116, American Mathematical Society, Providence.
Dauben, J.: 1979, *Georg Cantor*, Harvard Univ. Press, Cambridge, Mass.
Fitch, F. B.: 1948, 'An Extension of Basic Logic', *The Journal of Symbolic Logic* **13**, 95–106.
Fitch, F. B.: 1949, 'The Heine-Borel Theorem in Extended Basic Logic', *The Journal of Symbolic Logic* **14**, 9–15.
Fitch, F. B.: 1950, 'A Demonstrably Consistent Arithmetic – Part I', *The Journal of Symbolic Logic* **15**, 17–24.
Fitch, F. B.: 1951, 'A Demonstrably Consistent Arithmetic – Part II', *The Journal of Symbolic Logic* **16**, 121–124.
Lorenzen, P. and J. Myhill: 1959: 'Constructive Definition of Certain Analytic Sets of Numbers', *The Journal of Symbolic Logic* **24**, 37–49.
Schütte, K.: 1960, *Beweistheorie*, Springer, New York.

Dept. of Mathematics
SUNY at Buffalo
Buffalo, New York 14214
U.S.A.

EDITORIAL NOTE

In the future, *Synthese* will seek to publish from time to time special numbers devoted to Cognitive Science. We shall be happy to try to facilitate in this way the further growth of an important field which is close to the traditional concerns of our journal. Both new developments in the field and the general theoretical implications of such developments will be covered. We envisage both unrestricted numbers open to all contributions in the field of Cognitive Science and numbers with a more limited subject matter. Potential contributors are encouraged to submit their papers to the editors.

In connection with this new venture, the Editors are pleased to announce that Dr. Barry Richards has joined the Editorial Board of *Synthese*. Dr. Richards will act as a Special Consultant for work in the area of Cognitive Science, and papers in this area can be submitted to him also for possible publication in *Synthese*. The address of Dr. Richards is:

> The School of Epistemics
> University of Edinburgh
> 2 Buccleuch Place
> *Edinburgh EH8 9LW*
> Scotland

THE EDITORS

Synthese **60** (1984) 144.

ALEX ORENSTEIN

REFERENTIAL AND NONREFERENTIAL
SUBSTITUTIONAL QUANTIFIERS*

It is common to find philosophers claiming that it is possible to free the quantifiers – especially the particular (or so-called existential) quantifier – from questions of reference, existence, and ontology, by having recourse to what is now referred to as the substitutional interpretation of the quantifiers. Although there may be ontologically neutral uses of the substitutional interpretation, it is one of the goals of this paper to point out where this feature has been misconceived and to give equal due to uses of the substitutional account that have ontological import. I will examine the relation of the substitutional treatment of the quantifiers to the most important nonsubstitutional one, that of Tarski, with reference to recent claims that the difference between the two kinds is that the former lacks ontological significance.

Uncritical use of the somewhat entrenched Quinian terminology of "substitutional" versus "referential" quantification is most likely at the root of these mistaken views.[1] In his debates with Ruth Barcan-Marcus and his comments on Leśniewski, Quine has put forward the view that substitutional quantification has no referential force and thus lacks any ontological import.

Quantification ordinarily so-called is purely and simply the logical idiom of objective reference. When we reconstrue it in terms of substituted expressions rather than real values, we waive reference. We preserve distinctions between the true and false, as in truth-function logic itself, but we cease to depict the referential dimension. (Quine, 1966, p. 181)

Such is the course that has been favoured by Leśniewski and by Ruth Marcus. Its nonreferential orientation is seen in the fact that it makes no essential use of namehood. That is, additional quantifications could be explained whose variables are place-holders for words of any syntactical category. Substitutional quantification, as I call it, thus brings no way of distinguishing names from other vocabulary, nor any way of distinguishing between genuinely referential or value-taking variables and other place-holders. Ontology is thus meaningless for a theory whose only quantification is substitionally construed; meaningless, that is, insofar as the theory is considered in and of itself. The question of its ontology makes sense only relative to some translation of the theory into a background theory in which we use referential quantification. (Quine, 1969, pp. 63–64)

Given these views it is not surprising that some would claim to solve

Synthese **60** (1984) 145–157. 0039–7857/84/0602–0145 $01.30
© 1984 *by D. Reidel Publishing Company*

philosophical problems pertaining to quantifiers, reference, and ontology by simply switching to the substitutional approach. Thus, for example, Susan Haack in her book *Deviant Logic* suggests that a way of *tout court* taking care of the problem of particular-existential generalizations being theorems of logic is simply to switch to a substitutional interpretation on which, if Quine were right, no problems would arise.

The origin of these views can probably be traced back to a shallow analysis of the different interpretations of the quantifiers, i.e., a substitutional interpretation and a Tarskian account. The following is a somewhat oversimplified statement of these semantic conditions, which will nonetheless do for the points to be discussed in this paper. (I limit the discussion here to monadic predications.)

Tarskian Condition: '$(\exists x)Fx$' is true iff 'Fx' is satisfied by some object

Substitutional Condition S: '$(\exists x)Fx$' is true iff some substitution instance of '$(\exists x)Fx$' is true

(An instance of '$(\exists x)Fx$' results from replacing 'x' in 'Fx' by a constant of the category of singular terms, e.g., proper names.)

By restricting oneself to these conditions, one might come to believe that substitutional quantification has no ontological force because the explicans part of Condition S does not appeal to any word-object relation, e.g., satisfaction, assignments in domains, or reference. By contrast, the Tarskian condition appeals straightforwardly to objects (actually to sequences of objects) in virtue of the word-object (i.e., open sentence – sequence) relation of satisfaction. On the strength of this relation, Tarskian quantifications are correctly said to have referential force and ontological import.

However, one might proceed mistakenly on the basis of this comparison and regard the substitutional and referential distinction as providing a mutually exclusive classification; the error is reinforced if one equates the referential and the Tarskian.

Though such opinions are not uncommon, a little reflection will suffice to show that they are incorrect. An informal account of substitutional generalizations is that they are true if some (or all) of their instances are. Thus 'Something is a Siamese cat' would be true if 'Bouncer (my cat) is a Siamese cat' is true. A quantifier should be called substitutional when its truth conditions appeal to the truth of its instances. This, however, leaves open the question of how the instances themselves get to be true or false. On a natural account of the truth of instances – or the atomic sentences they depend upon – 'Bouncer is a Siamese cat', is true if and only if the object referred to by the subject is one of the objects to which the predicate applies. On such an explication of the truth of atomic sentences, the substitutionally interpreted generalization based on it would have referential force. The referential force and ontological import of '$(\exists x)$ (x is a Siamese cat)' would not be due to Substitutional Condition S but to the referential aspect of the truth condition for the instances it depends upon.

There are two relevant possibilities for interpreting atomic sentences. The first would be to follow the more natural course mentioned in the preceding paragraph and construe the atomic sentences as true if and only if the object denoted by the subject term were one of the objects to which the predicate applies. On this interpretation of the atomic sentence – plus the standard interpretation of truth-functional connectives – the instances would be such that substitutional generalizations would be *per accidens* a vehicle of reference. The reference achieved by asserting 'Some thing is a cat' would not be due directly to the substitutional truth conditions, but to the referential force of the instances appealed to in those substitutional conditions. If, as many, though not all, philosophers would assume, the referents of the terms in the atomic sentences are existents, the above particular generalization could and should be read existentially and as having ontological significance.[2] It is thus incorrect to say that what makes substitutional quantification different is its lack of referential, existential, or ontological force. Although, as we shall see, the substitutional account can be construed as ontologically neutral, whether or not it is so construed is not of the essence of the substitutional interpretation.

The second relevant possibility for giving truth conditions to our atomic sentences would be to do so without appealing to any such concepts as reference, designation, domains, etc. (i.e., without appealing to any word-object relations). One way of doing this is to say that an atomic sentence is true or holds if and only if it is a member of a given model or truth set of sentences. (The privileged given model or truth set would be the one that conforms to the actual world. This feature, however, is not appealed to in model or truth set approaches, which merely appeal to membership in a set of sentences.) The key point is that such an instance without referential force bestows no referential-ontological significance upon a substitutional generalization based on that instance. So where the instances are not construed referentially, the substitutional quantification has no such force.

   To illustrate this matter. let us consider two substitutional theorists, Hugues Leblanc and Ruth Barcan Marcus. Let me begin by contrasting these two theorists on the example 'Pegasus is a flying horse'. In a number of his papers as well as his book *Truth-Value Semantics*, Leblanc propounds a substitutional interpretation of the quantifiers. However, on singular terms he has at times held a referential view that is in effect very much like those of Quine and Russell, and what we above called natural (Leblanc, and Wisdom 1976, p. 148). According to Leblanc, in Leblanc and Wisdom, singular sentences such as the above are true only if the subject designates an existing object. A particular generalization derived from such a sentence is hence true only if something exists, so that his substitutionally construed first-order particular quantifier has the same existential force Quine and Russell would accord it. In contrast to Leblanc, consider Ruth Barcan Marcus's position in her 1962 paper 'Interpreting Quantification' (pp. 256–257). There she considered the Pegasus sentence above as true, presumably by divorcing its truth from any questions of reference. She sanctioned the particular generalization from that Pegasus sentence to the substitutionally construed: 'There is a true substitution instance of 'x is a winged horse''. The generalization is considered true in virtue of the truth of its instances and, unlike Leblanc's substitutional generalization, has no existential force.

   There is need for caution in using the terms 'substitutional' and 'referential' with regard to quantifiers and their interpretations. The use I am proposing is that the two are not exclusive of each other.[3] Leblanc's quantifiers have both substitutional and referential force. Quantifiers construed in a Tarskian fashion are nonsubstitutional and referential. Barcan Marcus's '$(\exists x)^s$ (x is a flying horse)' is substitutional and nonreferential. Let us regard an interpretation and an interpreted quantifier as substitutional when the explicans portion of the truth condition for the quantifier employs the relevant notion of a substitution instance. In constrast to a quantifier's being substitutional, a quantifier can be referential either because the interpretation of the quantifier is referential (Tarski) or the instances on which the generalization is based are interpreted referentially (Leblanc). The phrase 'referential interpretation' applies to either semantic conditions for quantifiers (Tarski) or conditions for atomic sentences (Leblanc). Of course in a trivial sense, the substitutional versus nonsubstitutional distinction is mutually exclusive.

There are at least three areas of problems where the substitutional interpretation has been applied in order to achieve ontological neutrality and thereby furnish solutions: (1) topics in free logic, (2) opaque constructions, and (3) quantifiers for grammatical categories other than singular terms.

In the first of these areas – free logic – the requirement is laid down that logic should be free of existence assumptions. In one sense, this means that logic should allow for individual constants that do not refer to any existents, i.e., vacuous singular terms, and that existential generalizations should not be logical truths, i.e., theorems.

To find a case where adopting the substitutional interpretation plus supplementing it with a nonexistential English reading of the particular quantifier was offered as a solution to a problem concerning vacuous singular terms, recall the Barcan Marcus proposal mentioned earlier. She maintained that if 'Pegasus is a flying horse' is true, then '$(\exists x)$ (is a flying horse)' is true as well, even though neither Pegasus nor any other flying horse exists. While there is nothing formally, i.e., purely logically, wrong with such a position, it is philosophically suspect. The position involves taking the instance as true with little if anything said as to how it gets to be true. This would be sanctioned formally on a substitutional semantics like Kripke's by taking the truth value of the atomic sentences for granted (Kripke, 1976, pp. 329–331). Another possibility is to say that the sentence holds in a given model set.

While this may be above reproach from the standpoint of formal logic and formal semantics, it appears to me more questionable than the philosophically and intuitively compelling view that regards such simple sentences as having a logical form that makes them true when the subject refers to an existing object of which the predicate is true, and false otherwise. Adopting a semantics that leaves the instances – atomic sentences – as true or false without informing us how, or by saying that they are members of a model set, i.e., that they are merely consistent with certain other sentences, should be disquieting. To follow such a course is to absolve oneself from answering the difficult question of what makes such sentences true or false. Indeed, to take the sentence as true simpliciter resembles truth by convention, and to make consistency the nature of truth smacks of the coherence theory of truth.

It is worthwhile to say a word here about the relation of interpretations of the quantifiers to natural-language readings for them. The semantical conditions (satisfaction or truth conditions) provide an account of how well-formed formulas governed by the '(x)' or the '$(\exists x)$' operations get to be true or false. Related to these conditions is the matter of precisely which English locutions are most appropriate to the different interpretations. Thus, where a Tarskian account of '$(\overset{T}{\exists} x)$' is given, especially by such authors as Quine, it is quite natural to read '$(\overset{T}{\exists} x)$' existentially in English as 'there exists'. It is also appropriate to give a special English reading to the substitutionally interpreted '$(\overset{S}{\exists} x)$' quantifier, viz., 'sometimes true' or 'for some x'. *The* substitutional reading provides no specific connotation of existence and thus has been used uncritically as the basis for spurious claims that substitutional quantification is ontologically neutral.

Like the substitutional interpretation – Condition S – the substitutional reading does not by itself guarantee that '$(\overset{S}{\exists} x)$' is non-referential and ontologically neutral. A substitutional reading of a quantifier can depend on instances that are referential. In such a situation there is little difference between the 'there exists' and the 'sometimes true' locutions. Indeed in Russell's writings, we find examples of the two readings used hand in hand (Russell, 1956, p. 46, and Whitehead and Russell, 1962, p. 15).

Susan Haack's suggested solution for the problem of what to do about existence theorems in logic is to adopt a substitutional reading so that the theorems in question will have no existential import (Haack, 1974, p. 143). It would then be unnecessary to revise a logical system so as to exclude particular generalizations being theorems. However, as noted in the preceding paragraph, merely switching to a substitutional 'sometimes true' reading of sentences such as '$(\overset{S}{\exists} x)$ $(Fx \vee \sim Fx)$' does not guarantee that '$(\overset{S}{\exists} x)$' has no referential-existential significance. The crucial question, granted that '$(\exists x)$' is interpreted substitutionally according to a condition such as S, is how the instances appealed to in S get their truth values. There are two possibilities. In the first, the instances are accounted for in the usual referential way, in which case '$(\exists x)\overset{S}{(Fx \vee \sim Fx)}$' has as much existential import as '$(\exists x)\overset{T}{(Fx \vee \sim Fx)}$' and should not be a theorem. The other possibility is that the truth values of the instances are accounted for without appealing to reference (word-object relations). There are at least two reasons why this will not do as a solution to the free logician's problem. In the first place, it is largely beside the point, since a system of logic where all the instances are construed nonreferentially is not the standard interpretation of quantificational logic even when it is construed substitutionally.

The second reason why this will not solve the problem even if we interpret all the instances nonreferentially involves a deeper parallel between substitutional and referential interpretations of the quantifiers. A more precise way of putting the free logician's point that particular-existential generalizations should not be theorems of logic is that a logical truth is a sentence that is true under every interpretation including the one involving the empty domain. Looking at the semantic condition below for a referentially interpreted quantifier, we see that a universal generalization will be true under the interpretation with the empty domain, i.e., the antecedent '$d \in D$' is false in the truth condition for the universal quantifier; however, an existential generalization will be false, i.e., the conjunct '$d \in D$' is false. The following is somewhat oversimplified, but will nonetheless do to highlight the points to be discussed here.

Referential Truth Conditions

val $((x)A) \overset{T}{=} T$ iff (d) $(d \in D \supset$ every relevant sequence of objects d with respect to 'x' leaves 'Ax' satisfied).

val $((\exists x)A) \overset{T}{=} T$ iff $(\exists d)$ $(d \in D$ and at least one relevant sequence of objects d with respect to 'x' leaves 'Ax' satisfied).

On a nonreferential substitutional account, a parallel problem arises. It is not a question of existence or of the empty domain, if the instances do not appeal to existent object or domains. It is a question of whether the classes of substituted constants can be empty. If there were no true instances, the substitutional universal generalization would be true, i.e., when the antecedent '$s \in L$' is false in the condition for '$\overset{S}{(x)}$' and the

particular generalization will be false, i.e., the conjunct '$s \in L$' is false
in the condition for '$(\exists\overset{s}{x})$'.

Substitutional Truth Conditions

val $((\overset{s}{x})A) = T$ iff $(s)\ (s \in L \supset$ val $(s/xA) = T)$ wherever x is free in A.

val $((\overset{s}{\exists x})A) = T$ iff $(\exists s)\ (s \in L$ and val $(s/xA) = T)$ [where s is a substituend in the language L].

When the substitution class is empty, we have an analogous problem to that posed by the empty domain for referential quantifiers. A possible example of this would arise for Quine's canonic notation, which contains individual variables but has no individual constants (names), if one interpreted the quantifiers for that notation substitutionally.

The second area where substitutional quantification has been put forward as a way of solving problems concerns quantifying into opaque constructions. Here as in the case of free logic, the problems are of two sorts. The first concerns applying such rules as existential generalization to opaque sentences. The second problem bears on whether certain sentences should be theorems of modal logic.

It has been argued by some that there is nothing problematic about inferences such as the following when the conclusion is construed substitutionally (Barcan Marcus, 1962, pp. 258–259, and Geach, 1972, pp. 139–146).

$$\frac{\Box \text{ (the evening star} = \text{the evening star)}}{\therefore (\exists\overset{s}{x})\ \Box\ (x = \text{the evening star)}}$$

If one takes the premise as true without further explanation and uses it as an instance that is the base for the substitutional generalization, then there is nothing formally wrong with such a move. It is probably faultless from the standpoint of formal logic and the formal semantics involved. But, as with the Pegasus example, the problem is philosophical. The question of the logical form of the instance and what determines its truth value have been sidestepped.

There has also been discussion as to whether the Barcan formula, $\Diamond(\exists x)Fx \supset (\exists x)\Diamond Fx$, is a truth of modal logic and ought to be a theorem. Barcan Marcus has taken the view that adopting a substitutional approach to the quantifiers will nullify the problem associated with the formula's being a logical truth. In "Interpreting Quantification" (1962, pp. 257–258), she suggested that the substitutional reading 'sometimes true' avoids the counterexample associated with reading the quantifiers in the formula existentially, i.e., if it is possible that there exists an F, then there exists something that is possibly an F. We noted earlier that merely switching readings or even interpretations of the quantifier does not guarantee that the substitutional reading is not also referential and existential; i.e., if the instances appealed to in the substitutional interpretation of the Barcan formula were construed referentially, then there might be no difference between the substitutional and the existential versions of the formula.

At a deeper level, there is an instructive similarity between the questions of whether existence sentences should be truths of first-order logic and whether the Barcan formula should be a truth of modal logic. The validity of a formula depends to a large extent on what the truth conditions for that formula are. To the extent that substitutional and nonsubstitutional truth conditions parallel each other (and a large amount of parallelism is to be expected – since they are two construals of the same formalism), sentences that are or are not logical truths should have their status preserved under the change in interpretation. Thus it is the semantics of modal logic, and not a superfical, informal reading of the quantifiers, that determines whether the Barcan formula is false under some interpretation. Kripke showed that it is false for what can be taken as a nonsubstitutional view of the quantifiers (Kripke, 1971, p. 67). Using parallel considerations but within a substitutional framework, Dunn has provided an analogous counterexample (Dunn, 1973, pp. 87–100).

The last area to be considered where the substitutional interpretation's purported ontological neutrality has been appealed to is in connection with quantifiers for diverse grammatical categories such as predicates, '$(\exists \overset{s}{\phi}) \phi a$', sentences, '$(\exists \overset{s}{p})\ (p \supset p)$', sentence operators, '$(\exists \overset{s}{f})\ (p f p)$', etc.

In an earlier paper (Orenstein, 1983), I argued that on a broad use of the concept of reference (which is in keeping with ordinary usage), we could speak of various parts of speech as referring. Thus both 'is human' and 'human' might be said to refer to men. This broad use of 'refers' is generic for various word-object relations, e.g., predicates applying, names denoting singularly, common nouns denoting multiply, etc. In this paper I will abide by the strict sense of referring, according to which only singular terms refer.

What then of the claim that substitutionally interpreted quantification for grammatical categories other than singular terms is ontologically neutral? For simplicity's sake I will confine the discussion to quantifying into predicate positions.

There are different ways in which these generalizations can be interpreted substitutionally when the relevant instances are interpreted referentially in the narrow or strict sense. Thus '$(\overset{s}{\exists}\phi)\phi a$', e.g., 'something is true of Alfred', is true if at least one instance of 'ϕa' is true (where the instance is obtained by substituting a predicate constant for the predicate variable 'ϕ'). This provides the substitutional interpretation of '$(\overset{s}{\exists}\phi)$'. What kind of truth condition would we provide for the instance 'Alfred is human' which could serve as a basis for the generalization. At least two alternatives with different ontological import come to mind. The ontologically more modest condition is that the sentence is true just in case the subject term refers to an object (individual) to which the predicate applies. The second interpretation requires treating all variables as having values which are referred to in the strict sense by the substituends for the variables. Hence 'is human', the substituend for 'ϕ', would refer here to the set of humans and purportedly in the strict and narrow sense. The instance is true because the object (individual) referred to by the subject term is a member of the set referred to by the predicate. Unlike the ontologically more modest account, this condition requires sets (or on an alternative view, properties), as well as the objects (individuals) referred to by the subject term. Following Quine, we could call this the "set theory in sheep's clothing" interpretation of higher-order logic. Though substitutional, such an account is also referential and has no ontological neutrality. It can be seen as a substitutional adaptation of Quine's view that to be is to be the value of a variable, and that different styles of quantification (e.g., as in higher-order logic) ontologically commit us to different types of entities. A similar exposition could be provided showing that substitutional and referential sentential quantification, e.g., '(p) $(p \supset p)$', would commit us to truth values or propositions.

Of course in Quine's own canonic notation, there is no place for any quantifiers save those for singular terms (all quantification is first-order quantification), and so only Tarskian satisfaction conditions are needed (Orenstein, 1977, pp. 67–68, 95–102). For Quine, higher-order logic (predicate variables and quantifiers) is set theory in disguise and '$(\exists \overset{s}{\phi})\phi a$' is misleading as to its logical form. He would find '$(\exists \overset{s}{\phi})\phi a$' written more appropriately as '$(\exists \overset{s}{x}) (a \in x)$' and the instance as 'Alfred $\in \{x|x$ is human$\}$'. For Quinians, a misleading feature of the interpretation above is that it mistakenly treats predicates as though they were singular terms. In other words predicate quantifications treat nonreferring (in the strict sense) positions as if they were referring (in the strict sense) positions. In fact if 'referential' in 'referential quantification' is taken in the narrow and strict sense (a relation between singular terms and their referents), then by definition there can be no referential quantification for nonreferring grammatical categories such as predicates, nouns, sentences, etc.

However, some – Prior, Williams, followers of Leśniewski, et al. – have found uses for '$(\exists \phi)\phi a$' and argue that it can be understood without appealing to sets of properties (Prior, 1971, and Williams, 1981). The only way I can see of doing this is to interpret the quantifiers substitutionally and then appeal to the ontologically more modest interpretation of the sentence 'Alfred is human', viz., Alfred is one of the objects to which 'is human' applies. Substitutional quantification on this view is not completely neutral, since it commits us ontologically to whatever objects that the predicate constants substituted for the variable are true of i.e., apply to. '$(\exists \overset{s}{\phi})\phi a$' does have existential import, but only for objects of the ontological category to which the singular term strictly refers, and of which the predicate constant is true. Thus, if the predicate constant were 'is human', commitment would be to a concrete object, while if the predicate constant were 'is odd', commitment would be to an abstract object.

Thus the claim that providing a substitutional interpretation for predicate quantifiers is ontologically neutral runs into difficulties. A truly ontologically neutral interpretation of '$(\exists \overset{s}{\phi})\phi a$' would interpret the generalization substitutionally, and then interpret the relevant instances without appealing to any word-object relation. However, such an interpretation would be nonstandard. One is left here with the difficulty discussed earlier of saying how an instance such as 'Alfred is human' that could be the basis for a substitutional generalization has a truth value without appealing to any word-object relation.

NOTES

* I would like to express my thanks for a grant from the Research Foundation of the City University of New York to aid in preparing this paper. I would also like to thank Helen Lauer, Pamela Noguerola, Deborah Sampson, and Leah Savion for their suggestions on earlier versions of this paper, and Hugues Leblanc for many helpful comments on the present one.

[1] Quine also uses the term "objectual quantification", and as interchangable with "referential quantification". I have avoided the former locution in this paper, and in Orenstein 1983 I have suggested different uses for the two terms.

[2] Among the philosophers excluded here are Meinongians who appeal to referents which subsist but don't exist.

[3] Professor Quine clarified his own position in correspondence with me: "You write that 'Quine's introduction of the terms "referential" and "substitutional" quantification suggests that the two kinds are mutually exclusive.' It shouldn't suggest that. They overlap. Quantification in elementary number theory is referential and substitutional; there is no difference when everything quantified over has a designator. See *Roots of Reference*, p. 114, lines 21–24; *Ways of Paradox*, enlarged edition, pp. 318–320; *Philosophy of Logic*, page 92, fourth line from bottom. Actually my view is the one you propose." (Quine is here referring to the combination of the substitutional and referential views for first-order cases discussed in this essay.)

REFERENCES

Dunn, J. Michael: 1973, 'A Truth Value Semantics for Modal Systems', in H. Leblanc (ed.), *Truth, Syntax and Modality*, North Holland, Amsterdam, pp. 177–185.

Geach, Peter: 1972, *Logic Matters*, Basil Blackwell, Oxford.

Haack, Susan: 1974, *Deviant Logic*, Cambridge University Press, London.

Kripke, Saul: 1971, 'Semantical Considerations on Modal Logic', in Leonard Linsky (ed.), *Reference and Modality*, Oxford University Press, Oxford, pp. 63–72.

Kripke, Saul: 1976, 'Is There a Problem About Substitutional Quantification?' in G. Evans and J. McDowell (eds.), *Truth and Meaning*, Oxford University Press, Oxford, pp. 325–419.

Leblanc, H.: 1976, *Truth-Value Semantics*, North Holland, Amsterdam.

Leblanc, H.: 1971, 'Truth-Value Semantics for a Logic of Existence', *Notre Dame Journal of Formal Logic* 12, 153–168.

Marcus, Ruth Barcan: 1962, 'Interpreting Quantification', *Inquiry* 5, 252–259.

Marcus, Ruth Barcan: 1976, 'Dispensing with Possibilia', *Proceedings and Addresses of the American Philosophical Association* 49.

Orenstein, Alex: 1977, *Willard Van Orman Quine*, Twayne Publishers, Boston.

Orenstein, Alex: 1978, *Existence and the Particular Quantifier*, Temple University Press, Philadelphia.

Orenstein, Alex: 1983, 'Towards a Philosophical Classification of Quantifiers', in Alex

Orenstein and Raphael Stern (eds.), *Developments in Semantics*, Haven Publications, New York.

Prior, Arthur: 1971, *Objects of Thought*, Clarendon Press, Oxford.

Quine, Willard Van Orman: 1966, *The Ways of Paradox and Other Essays*, Random House, New York.

Quine, Willard Van Orman: 1969, *Ontological Relativity and Other Essays*, Columbia University Press, New York.

Quine, Willard Van Orman: 1973, *The Roots of Reference*, Open Court, La Salle.

Leblanc, H. and W. A. Wisdom: 1976, *Deductive Logic*, Allyn & Bacon, Boston.

Russell, Bertrand: 1956, *Logic and Knowledge*, Macmillan, New York.

Whitehead, A. N. and B. Russell: 1962, *Principia Mathematica*, Cambridge University Press, Cambridge.

Williams, C. J. F.: 1981, *What is Existence?* Clarendon Press, Oxford.

Dept. of Philosophy
Queens College and the Graduate School
The City University of New York
33 West 42nd St.
New York, NY 10036
U.S.A.

WILFRIED SIEG

FOUNDATIONS FOR ANALYSIS AND PROOF THEORY*

> Douter de tout ou tout croire, ce sont deux solutions
> également commodes qui l'une et l'autre nous dis-
> pensent de réfléchir.
>
> H. Poincaré

INTRODUCTION

The title of my paper indicates that I plan to write about foundations for analysis and about proof theory; however, I do not intend to write about *the* foundations for analysis and thus not about analysis viewed from the vantage point of any "school" in the philosophy of mathematics. Rather, I shall report on some mathematical and proof-theoretic investigations which provide material for (philosophical) reflection. These investigations concern the informal mathematical theory of the continuum, on the one hand, and formal systems in which parts of the informal theory can be developed, on the other. The proof-theoretic results of greatest interest for my purposes are of the following form:

> for each F in a class of sentences, F is provable in T if and
> only if F is provable in T^*,

where T is a classical set-theoretic system for analysis and T^* a constructive theory. In that case, T is called REDUCIBLE TO T^*, as the principles of T^* are more elementary and more restricted.

I also want to emphasize from the outset that I do not think such reductions are needed to "justify" the practice of classical mathematics. Paul Bernays remarked in 1930, and I cannot but agree with him today, that

the current discussion on the foundations of mathematics does not arise from any emergency within mathematics itself. Mathematics is in a perfectly satisfactory state of certainty concerning its methods.[1]

He viewed the critical issue in the discussion as follows:

Synthese **60** (1984) 159–200.   0039–7857/84/0602–0159 $04.20
© 1984 *by D. Reidel Publishing Company*

The problems, the difficulties, and the differences of opinion begin only, when one inquires not simply after the mathematical facts, but after the grounds of mathematical knowledge and after the delimitations of mathematics.[1]

These inquiries of a reflective, philosophical character continue to have profound interest. Their fruitful pursuit, however, must be informed by detailed mathematical and metamathematical work. Before either accepting or rejecting portions of mathematics on philosophical grounds, we have to analyze more carefully and understand more adequately this central part of our intellectual experience.

But what can proof theory, you may wonder, contribute to such a better understanding? Is it not most intimately connected with a particular foundational program, namely, Hilbert's? And is that not inspired by a crude formalism which, if it managed to survive Frege's criticism by some miracle, is certainly dead since Gödel proved his incompleteness theorems? These questions will be answered implicitly in this paper. I shall explicitly discuss a modification of Hilbert's program which has been explained by Bernays in numerous writings. In that modified form the program provides a coherent framework for foundational research focusing on two complementary goals: to determine (i) which principles are used and needed in a particular branch of mathematics, primarily in classical analysis, and (ii) which constructive, and possibly more evident, principles suffice for a reduction of theories arising from (i).

Recently, most interesting discoveries related to (i) and (ii) have been made. With regard to (i) it has been found that the bulk of classical analysis can be carried out in conservative extensions of elementary number theory; with regard to (ii) it has been shown that strong impredicative classical theories for analysis can be reduced to constructively acceptable ones. I will describe these results in some detail; first, however, I will sketch their historical and systematic background. Before beginning with this sketch, let me emphasize again that I want to report on work which provides, in my view, material for philosophical reflection. I will not put forward philosophical theses here, but point out objective mathematical and metamathematical relations. Perhaps that can better than any direct argument contribute to the insight, that the (exclusive) alternative between "constructivistic" and "platonistic" foundations is "a logically inadmissible application of the tertium non datur". (That is how Zermelo characterized the alternative between Brouwer's intuitionism and Hilbert's formalism.)

PART 1 – FOUNDATIONS FOR CLASSICAL ANALYSIS

In this first part of my paper I shall outline very schematically (the development towards) the arithmetization of analysis in 19th century mathematics. The main reason for presenting this historical sketch is to point beyond the set-theoretic paradoxes, which occupy such a pivotal place in the modern discussion on the foundations of mathematics, to the longer standing problem of providing a sound basis for analysis. A reduction or a radical restriction to the natural numbers was viewed by many mathematicians as a way of solving this problem in a mathematically and philosophically satisfactory manner. It seems to me that "the" arithmetization of analysis was the common concern of men like Dedekind, Cantor, and Weierstrass, Frege and Russell, Kronecker and Brouwer, Hilbert and Zermelo. This last statement may be more plausible when one observes that none of these mathematicians appealed in their foundational work to other than narrowly arithmetic and (or) logical, set-theoretic concepts; in particular, an appeal to geometric concepts was never contemplated.[2]

The very introduction of irrational numbers into analysis, however, is geometrically motivated; the notion of an arbitrary cut of rationals in Dedekind's work is suggested by considering completely arbitrary divisions of a straight line into two segments. That notion, or that of an arbitrary subset of natural numbers, is involved in the arithmetization of analysis as given by Dedekind and Weierstrass. This is one point I want to emphasize already now. Another point has to do with the development of Hilbert's views on how to give consistency proofs for analysis: from adapting the Dedekind-Weierstrass methods to reflecting on analysis as codified in a formal theory by elementary mathematical means. He hoped to achieve in this radically new way a more thoroughgoing, stricter arithmetization, namely, an arithmetization satisfying the finitist restrictions of Kronecker.

1.1 Arithmetization of Analysis

One may go back to Bishop Berkeley's "The Analyst, or a Discourse Addressed to an Infidel Mathematician" (1734) for a vigorous and biting attack on inconsistencies in the early calculus. Even a hundred years later fundamental parts of analysis were still obscure, not only to philosophers and theologians but to very gifted mathematicians. N. H.

Abel vowed in a letter of March 29, 1826 to C. Hansteen

to apply all my energy to bring a little more clarity into the surprising obscurity one finds
undoubtedly in analysis today. It lacks all plan and unity . . . . the worst is that it has not at
all been treated with rigor. There are only a very few theorems in advanced analysis which
are proved with complete rigor.[3]

Abel's complaint was certainly justified: basic notions of analysis
(continuity for example) were vague; geometric images often took the
place of strict proof. Indeed, the central concepts of derivative and
integral had neither a proper definition nor a clear range of applicability. So, it is not surprising that eminent mathematicians of the
19th century attempted to secure a rigorous basis for analysis.

Euler and Lagrange had already indicated a direction for the search
of such a basis: analysis was to be built exclusively on arithmetic
concepts. The tendency to consider the natural numbers as the ultimate
basis in mathematics was undoubtedly strengthened by the discovery of
Non-Euclidean geometry. Gauss drew philosophical consequences
when distinguishing (in a letter of 1817) arithmetic from geometry by
observing that only the former was *a priori*. Thirteen years later he
wrote to Bessel:

Geometry has, according to my deepest convictions, a completely different relation to our
knowledge a priori than pure arithmetic; our knowledge of the former is indeed lacking in
that complete conviction of its necessity (and thus also of its absolute truth), which is
characteristic of the *latter*; we have to admit in humility, that if number is *merely* the
product of our mind, space has also outside of our mind a reality, to which we cannot a
priori completely prescribe its laws.[4]

Dirichlet, the successor of Gauss in Göttingen, claimed repeatedly
that *any* theorem of analysis could be formulated as a theorem
concerning the natural numbers. This is related by Dedekind.[5] Dirichlet's claim must have been based on two convictions, namely, that the
advanced notions of analysis can be defined arithmetically (which had
already been achieved by Cauchy for continuity), and that the real
numbers can be "constructed" from the natural numbers in a purely
arithmetic fashion. The construction of the set R of reals from the set N
of natural numbers is for my discussion most significant; in particular
the step from Q, the rational numbers, to R. This step was taken in
different, but equivalent ways by Weierstrass, Cantor, and Dedekind.

Dedekind's construction, on which I shall concentrate, was conceived in 1858 and published 14 years later in "Stetigkeit und irrationale

Zahlen". It is thoroughly geometrically *motivated*! The observation that each point of a straight line partitions that line into segments S_1 and S_2, such that each point of S_1 lies to the left of each point of S_2, led Dedekind to the formulation of a principle expressing the "essence of continuity". It is the converse of the above observation:

If all points of the straight line fall into two classes, such that each point of the first class lies to the left of each point of the second class, then there exists one and only one point producing this partition of all points into two classes, this cutting (Zerschneidung) of the straight line into two segments.[6]

The field of rational numbers Q, whose construction from N is assumed in that essay, has to be completed to a "continuous manifold" (stetiges Gebiet), if one wants to pursue all properties of the straight line arithmetically. But how can this be done? By carrying the geometric continuity principle over to arithmetic and letting the real numbers just *be* cuts of rationals, i.e., partitions of Q into two nonempty classes A_1 and A_2 with the characteristic property that each element of A_1 is smaller than each element of A_2.[7] Dedekind proves then that the system of all cuts of rationals is a continuous manifold: each cut of real numbers is already determined by a unique cut of rationals (Satz IV). R is thus, Dedekind writes in a letter to Lipschitz, "das denkbar vollständigste Grössen-Gebiet".[8]

The continuity principle is not only motivated from a geometric standpoint; it is indeed intrinsic to the practice of analysis. Dedekind points that out by showing that Satz IV is equivalent to two main theorems. Let me mention the first of them; it asserts that any monotonically increasing, but bounded function (from R to R) approaches a limit. Dedekind says in the introduction, when describing his reasons for developing a theory of irrationals, that in proofs of this theorem he always had to appeal to geometric evidence. Another theorem which depends crucially on the continuity principle and which is obvious when we think of continuous manifolds is this: if g is a continuous function on the closed interval $[0,1]$ with $g(0) < 0$ and $g(1) > 0$, then g vanishes at some point in the interval.[9] (Note here that the principle does what it is supposed to do; it allows us to prove in "an arithmetic manner" geometric properties characteristic of continuous manifolds.)

Dedekind viewed a cut as a purely arithmetic phenomenon (eine rein arithmetische Erscheinung).[10] This seems to be correct, however, only if parts of set theory are adjoined to arithmetic (as it is understood now), or if arithmetic is a part of logic and the needed set-theoretical principles are available as logical ones (as Dedekind believed).[11] Dedekind formulated such principles explicitly in his other foundational essay "Was sind und was sollen die Zahlen?" He took, for example, as a logical law that the extension of any predicate is a "system", his term for set, manifold, or totality. A system is completely determined, if for each thing (which for Dedekind is any object of our thinking, in particular, systems are things) it is determined whether or not it is an element of the system. Dedekind remarks in a footnote that it is quite irrelevant for his development of the general laws of systems, whether their complete determination can be decided by us. This remark is directed against Kronecker who had earlier on formulated decidability conditions. I shall come back to this later; but let me note that such restrictions form one important reason for Kronecker's rejecting the general notion of irrational number.

Using such logical principles, and an assumption I shall formulate at the beginning of 1.2, Dedekind showed in that essay how the natural numbers can be defined and characterized up to isomorphism. What a masterpiece of mathematical development and conceptual analysis! (For an appreciation of the latter it is enlightening to read Dedekind's letter to Keferstein, translated and published in [van Heijenoort].) The step from $\mathbb{N}$ to $\mathbb{Q}$, which is still missing for a complete "arithmetical" construction of $\mathbb{R}$, can be taken by making use of Kronecker's "Über den Zahlbegriff".[12] In that paper, published one year before Dedekind's essay in 1887, Kronecker showed how to eliminate from arithmetic systematically all those notions which he considered to be foreign to it, namely, the negative and rational numbers, the real and imaginary algebraic numbers.[12] Kronecker's "elimination" of negative and rational numbers can easily be turned into their "creation" in terms of equivalence classes of pairs of natural numbers, and that is quite in Dedekind's spirit.

The view that a strict and satisfactory arithmetization of analysis had been achieved was widely shared among mathematicians by the end of the 19th century, in spite of Kronecker's opposition to the very introduction of the general notion of irrational number. This conviction is most vividly expressed in Poincaré's remarks to the Second International Congress of Mathematicians (Paris, 1900):

Today, in analysis there are only natural numbers or finite or infinite systems of natural numbers . . . . Mathematics, as one says, has been arithmetized. . . . In today's analysis, if one cares to be rigorous, there are only syllogisms and appeals to that intuition of pure numbers which alone cannot deceive us. One can say that absolute rigor has been achieved today.[13]

Poincaré, in contrast to Dedekind, took the natural numbers as fundamental mathematical objects; and it may be correct that intuition of pure numbers cannot deceive us. However, in the arithmetization of analysis infinite systems are used. That concept, even when restricted by "of natural numbers", turned out to be more problematic than either Dedekind or Poincaré had thought.

1.2. A set (of new and old problems)

The final step of Dedekind's arithmetization of analysis (or should one say "reduction to logic"?) involved an argument for the existence of infinite systems; it made crucial use of the assumption that the "totality of all things which can be an object of my thinking"[14] is a system. Cantor told Dedekind in a letter of July 28, 1899, that this assumption leads to a contradiction. In his own investigations, Cantor had discovered the necessity of distinguishing between two types of multiplicities (systems) of things:

For a multiplicity can be such that the assumption that all its elements "are together" leads to a contradiction, so that it is impossible to conceive of the multiplicity as a unity, as "one finished things". Such "multiplicities" I call absolutely infinite or inconsistent multiplicities.[15]

And he added:

As we can readily see, the "totality of everything thinkable", for example, is such a multiplicity. . . .[15]

Which objective criterion can ensure the consistency of a multiplicity and "thus" its existence as a set? Hilbert who had been informed of the Cantorian difficulties proposed a solution at least for the multiplicity of real numbers.[16] In a paper which he entitled "Über den Zahlbegriff" (undoubtedly in polemical allusion to Kronecker's essay of the same title[17]) he presented a categorical axiomatization for $\mathbb{R}$. He claimed that a consistency proof for this axiom system could be given by a suitable modification of Dedekind's (or Weierstrass's) methods in the theory of irrational numbers. In Hilbert's view, the proof of the consistency would at the same time be

the proof of the existence of the totality of real numbers or – to use G. Cantor's terminology – the proof of the fact, that the system of real numbers is a consistent (finished) set.[18]

The same point was made in his address to the Paris Congress of Mathematicians, when discussing the second of his famous problems. Actually, he went further and claimed that the existence of Cantor's higher number classes and of the alephs could be proved in a similar manner. As in the earlier paper, he mentioned that the existence of the totality of all number classes or all alephs cannot be established (in this way): these multiplicities are nonconsistent, nonfinished.[19]

By 1904, Hilbert had given up the view that the consistency of his axiom system for R could be established by means of Dedekind's methods, suitably modified. The plausible reason for this change of mind is the fact that in the meantime more difficulties had arisen in set theory; most importantly, Zermelo and Russell had pointed out the contradiction derived from the set of all sets not containing themselves as elements.[20] Hilbert gave up the particular way in which he had hoped to prove consistency, but he did not give up the goal. The thread of this development is taken up again later. At this point I just mention that Hilbert in [1904] proposed a simultaneous development of logic and arithmetic. As logic was understood in the broad sense of Dedekind and Frege to contain set-theoretical principles, his proposal included (or rather called for) an axiomatization of such principles, which could serve as the basis for mathematics. Within four years, Zermelo and Russell had formulated theories which were suitable for that task and in which the known paradoxes could not be derived, at least not by the usual arguments.[21]

Around 1900, Zermelo had begun, under the influence of Hilbert, to turn his attention to the foundations of mathematics and in particular to the basic problems of Cantorian set theory.[22] In his paper 'Untersuchungen über die Grundlagen der Mengenlehre I' he undertook to isolate the set-formation principles central for the development of set theory and thus, in Zermelo's view, of mathematics. For this he analyzed Cantor's and Dedekind's work very closely. The somewhat cautious and experimental attitude in which this enterprise was pursued is best described by Zermelo himself:

In solving the problem [of establishing the foundations for set theory] we must, on the one hand, restrict these principles sufficiently to exclude all contradictions, and, on the other, take them sufficiently wide to retain all that is valuable in this theory.[23]

It is all too well-known which principles were taken by Zermelo as starting-points for his axiom system. Quite in the Hilbertian spirit, Zermelo considered a consistency proof for his axioms to be "very essential".[24] There was, however, a further problem, the more philosophical question "about the origin of these set-theoretic principles and the extent to which they are valid."[25] A mathematically and conceptually convincing answer to this question was given more than twenty years later in Zermelo's essay 'Über Grenzzahlen und Mengenbereiche'; segments of the cumulative hierarchy are recognized as the domains, in which the axioms (expanded by foundation and replacement) are valid. One may put this differently by saying that the ZF-axioms formulate the principles underlying the construction of the cumulative hierarchy.

Russell started out with a detailed analysis of the paradoxes and proposed in [1908] as a final solution his theory of ramified types. This was motivated by broad philosophical considerations.[26] Yet for the development of mathematics within that logical framework, he was forced to make the assumption of the axiom of reducibility.[27]

This assumption seems to be the essence of the usual assumption of classes; at any rate, it retains as much of classes as we have any use for, and little enough to avoid the contradictions which a less grudging admission of classes is apt to entail.[28]

The pragmatic use of the axiomatic method (in a stricter form than Zermelo's) comes to the fore even more strikingly in *Principia Mathematica*. In that work, Whitehead and Russell engaged in a quasi-empirical study. They wanted to demonstrate that contemporary mathematics could be *formally* developed in the theory of types (with the axiom of reducibility). The spirit in which the work was undertaken is worthwhile recalling. In constructing a deductive system for mathematics, they say in the preface to the first edition of *Principia Mathematica*, one has to perform two concurrent tasks: to analyse which principles are actually used and to rebuild mathematics on the basis of those principles. This is a perfectly standard account of axiomatic work in mathematics and indeed any other sufficiently developed subject. It is preceded by a rather provocative and perhaps startling claim:

We have, however, avoided both controversy and general philosophy, and made our statements dogmatic in form. The justification for this is that the chief reason in favour of any theory on the principles of mathematics must always be inductive, i.e. it must lie in the fact that the theory in question enables us to deduce ordinary mathematics. In mathematics, the greatest degree of self-evidence is usually not to be found quite at the beginning, but at some later point; hence the early deductions, until they reach this point, give reasons rather for believing the premises because true consequences follow from them, than for believing the consequences because they follow from the premises.[29]

Whether Whitehead and Russell succeeded in avoiding controversy with their logical work, I leave for the reader to judge; they definitely did not with their statement that "the chief reason in favour of *any* theory on the principles of mathematics must *always* be inductive". There are not only convincing conceptual analyses of mathematical principles (e.g., Dedekind's for the so-called Peano-axioms or Zermelo's for the ZF-axioms), but there are also areas of "ordinary mathematics" where the "original data" are quite dubious (e.g., infinitesimals in the early calculus). To put it squarely, the creative and critical function of fundamental theories is not appreciated here.

It is almost amusing to consider Whitehead and Russell's maxim for arguing in favour of a basic theory side by side with their view that the (Dedekind, Cantor) theory of irrational numbers is undoubtedly a part of ordinary mathematics![30] (And thus, one presumes, belongs to the "original data" to be accounted for.) The controversy surrounding the irrationals and the set-theoretic principles used in their definition, after all, had not been resolved. It was also connected with broader, more philosophical issues concerning the nature of mathematics. The public debate between mathematicians flared up more intensely again after Zermelo's proof of the well-ordering theorem by means of the axiom of choice [Zermelo, 1904]. The reluctance of mathematicians to accept the result was founded (mostly) in the strikingly nonconstructive, purely existential character of the choice principle. Though it is obviously true for Zermelo's notion of set and his broad subset notion,[31] it loses its evidence when only definable subsets of a given domain are considered. In the famous exchange of letters between Borel, Baire, Lebesgue, and Hadamard concerning Zermelo's proof, the first three considered it to be crucial for a secure foundation of analysis to admit only definable subsets of natural numbers.[32] Lebesgue associated a negative answer to the question, Can one prove the existence of a mathematical entity without defining it? explicitly with Kronecker. Baire seemed to suggest that all infinite totalities should be banished from mathematics. He writes at the end of his letter:

And, finally, despite appearances, everything must be reducible to the finite.[33]

Baire's statement is certainly in the tradition of Kronecker. The latter had rejected from the start the set-theoretic treatment of the infinite (contra Cantor), the introduction of irrational numbers (contra Weierstrass), and nonconstructive methods in algebra (contra Dedekind).[34] Kronecker had been at the center of the controversy in the 1870s and 1880s.[35] Unfortunately, Kronecker published only one paper concerned with foundational matters, his 'Über den Zahlbegriff'. In that paper he formulates and pursues a radically eliminative program, as described in 1.1.[36] Arithmetic is for him, as it was for Gauss, the "Queen of Mathematics" and comprises all mathematical disciplines with the exception of geometry and mechanics. Algebra and analysis in particular fall under it. Absolutely fundamental are the natural numbers;[37] he believes, that

we shall succeed in the future to "arithmetize" the whole content of all these mathematical disciplines [which fall under arithmetic]; i.e. to base it [the whole content] on the concept of number taken in its most narrow sense, and thus to strip away the modifications and extensions of this concept, which have been brought about in most cases by applications in geometry and mechanics.[38]

(In a footnote, referring to "extension of this concept" in the above quotation, Kronecker makes clear that he has in mind "in particular the addition of the irrational and continuous magnitudes".) Immediately after this passage with its call for strict arithmetization, Kronecker points to the *difference in principle* between arithmetic (in his sense) on the one hand and geometry and mechanics on the other: only the former is *a priori*. As support for his position, he describes Gauss's view expressed in the letter to Bessel, which I quoted when discussing the arithmetization of analysis. (As a matter of fact, that text is given by Kronecker in a footnote.)

The restriction to the natural numbers as the legitimate objects of (Kronecker's) arithmetic is accompanied by a restriction of methods. The latter is hinted at in the effective treatment of roots of algebraic equations in 'Über den Zahlbegriff';[39] it is discussed explicitly and in general terms in Hensel's foreword to Kronecker's 'Vorlesungen zur Zahlentheorie'. As the general points are so characteristic for a (very restricted) constructivist approach to mathematics, let me quote these remarks.

He [Kronecker] thought that . . . any definition should be formulated in such a way, that one can find out in a finite number of trials, whether it applies to a given magnitude or not. Similarly, a proof of existence of a magnitude can only be viewed as completely rigorous, if it contains a method which allows us to find the magnitude whose existence has been claimed.[40]

It is in this critical, constructive tradition within mathematics that Brouwer and his intuitionism can be seen.[41] Brouwer, however, went beyond restricting classical mathematics: he introduced new notions (e.g., that of a choice sequence), new methods (e.g., bar induction), and a constructive treatment of infinitary objects (ordinals or well-founded trees). This led actually to a development of analysis which is in conflict with classical logic.[42] Brouwer attempted to rebuild on an intuitionist basis parts of classical mathematics, in particular, parts of set theory and analysis;[43] but he had no qualms about discarding what did not measure up to intuitionistic principles. And that seemed to be large and substantial parts of the classical theory.

1.3 *Hilbert's Program*

Others were not prepared to follow Brouwer or, earlier on, Kronecker in restricting mathematical methods and set theory (more than necessary to avoid contradictions). Most outspoken among those was Hilbert. Bernays is quoted in C. Reid's Hilbert biography as saying:

Under the influence of the discovery of the antinomies in set theory, Hilbert temporarily thought that Kronecker had probably been right there. [i.e., right in insisting on restricted methods.] But soon he changed his mind. Now it became his goal, one might say, to do battle with Kronecker with his own weapons of finiteness by means of a modified conception of mathematics . . . .[44]

In his Heidelberg talk of 1904, Hilbert sketched what he considered to be a refutation (sachliche Widerlegung)[45] of Kronecker's viewpoint on foundations. The ultimate goal of his proposal was the same as the one formulated in 'Über den Zahlbegriff' and his Paris address, namely, to establish the existence of the set of natural and real numbers and of the Cantorian alephs;[46] and that was to be achieved, as before, by consistency proofs. In this talk, however, Hilbert indicates how such proofs might be given without presupposing basic logical, set-theoretic notions. Such presuppositions seem to him to be problematic now for the foundation of arithmetic, as they in turn use basic arithmetic concepts. So he proposes a "simultaneous development of the laws of logic [understood in Dedekind's broad sense] and arithmetic".[47] The development is to be given in a stricter, more formal way so that proofs can be viewed as finite mathematical structures (endliche mathematische Gebilde). The new task is to show by elementary mathematical means that such proofs cannot lead to a contradiction.[48] However vague, provisional, and confused these suggestions may have been, they foreshadowed aspects of Hilbert's proof-theoretic program of the twenties.

It is to be noted here that Hilbert was not opposed to constructive tendencies in mathematics; indeed, he later claimed that only his program does justice to them "as far as they are natural".[49] Furthermore, he admitted that "Kronecker's criticism of the usual way of dealing with the infinite was partly justified".[50] From papers written by Bernays in 1921 and 1922 we can plausibly infer that Hilbert viewed his program then as a way of mediating between two conflicting doctrines, namely, the classical set theoretic and the (intuitionistic) constructive one.[51] How could that be achieved? – By an epistemological reduction! That is most clearly formulated in [Bernays, 1922A, pp. 10–11] (and in a more refined form in [Bernays, 1930, pp. 54–55] and in [Hilbert/Bernays, Bd. I, pp. 42–44]). The axiomatic theories, which had been given for number theory, analysis, and set theory, and their implicit existential assumptions had to be justified by appealing only to the most primitive intuitive cognitions (primitiv anschauliche Erkenntnisse [Bernays, 1922A, p. 11]). This justification was not to guarantee the truth of those assumptions, but the consistent development of the axiomatic theories and thus the truth of elementary statements.[52] (This will be explained in greater detail below.) As the consistency problem could be formulated in an elementary mathematical way, there was at least a chance of solving it by constructive means. Those means were included in the most elementary part of number theory. Hilbert called this part of mathematics FINITIST and believed that it coincided essentially with what Kronecker and Brouwer (had) accepted.[53] Let me describe more precisely how such a justification was to be given. (The particular way of giving this description is due to Kreisel; see for example [Kreisel, 1968].)

For Russell, the work in *Principia Mathematica* established the reduction of mathematics to logic; for Hilbert, it simply showed that (parts of) mathematics can be developed within a particular formal system S. Basic assumptions are formulated as axioms, and logical inferences are specified as rules; thus, mathematical reasoning, as far as its results are concerned, can be replaced by manipulating symbols according to fixed rules.[54] Hilbert made the crucial observation that such a formal system can be described from a metamathematical point of view in a finitist way. In particular one can express finitistically that a certain configuration b constitutes a proof of a formula F in the system S; briefly,

$$\mathrm{Pr}_S(b, F).$$

And this notion satisfies Kronecker's methodological desideratum, as one can decide in a finite number of steps whether a given configuration is or is not an S-proof of F.

Using the proof predicate Pr_S one can formulate what Kreisel calls ADEQUACY CRITERIA for S with respect to finitist mathematics $\mathscr{F}$. The first criterion requires that finitist mathematics can be developed in S. For the formulation of the criterion let $\pi(\tau)$ be a finitist mapping associating derivations (formulas) in S with statements $\mathscr{A}$ of $\mathscr{F}$:[55]

(Adeq 1) $\mathscr{A} \rightarrow \mathrm{Pr}_S(\pi(\mathscr{A}), \tau(\mathscr{A}))$.

This condition seemed to be satisfied by those theories in which Hilbert was most interested; in set theory, for example, all of elementary number theory could be developed. The second adequacy criterion is crucial. It expresses in finitist terms that S proves only correct $\mathscr{F}$-statements;[56] more concisely,

(Adeq 2) $\mathrm{Pr}_S(b, \tau(\mathscr{A})) \rightarrow \mathscr{A}$.

Adeq 2 is a proof-theoretic REFLECTION PRINCIPLE, and it is equivalent to the consistency statement for S.[57] The central question for work on Hilbert's program was consequently: *can one establish the reflection principle (or the consistency statement) for S by finitist means?* – Notice, as Hilbert did[58], that a finitist consistency proof for S would yield a method to transform any S-proof of $\tau(\mathscr{A})$ into a finitist proof of $\mathscr{A}$. In Hilbert's own language, a finitist consistency proof would allow us to eliminate ideal statements from the proof of real statements. (We shall see in 2.2 below how "literally" this can be achieved by Gentzen's method.)

Remark. Hilbert thought that finitist mathematics was a philosophically unproblematic and absolutely basic part of mathematics; by his program he hoped to have separated the foundational questions in mathematics from general philosophy, and to have formulated them in such a way that they were amenable to a *final*, mathematical solution. Proof theory was to settle these questions – once and for all! "Ich möchte nämlich die Grundlagenfragen der Mathematik als solche endgültig aus der Welt schaffen . . . ."[59]

The central question of the program initially allowed for some successful answers. Ackermann and von Neumann gave consistency proofs for elementary number theory, or so they thought. A consistency proof for analysis seemed to be within reach.[60] Elementary number theory was believed to be complete.[61] These speculations were limited severely by Gödel's incompleteness theorems. The second theorem showed immediately that the results of Ackermann and von Neumann were of restricted scope and applied only to a part of elementary number theory.[62] It implied furthermore a general restriction on consistency proofs for sufficiently strong formal theories S: consistency proofs for S cannot be given by means formalizable in S. Gödel noticed, however, at the end of his paper that this result does not contradict Hilbert's formalistic standpoint:

For the latter presupposes only the existence of a consistency proof carried out by finitist methods, and it is conceivable that there might be finitist proofs which cannot be represented in P . . . .[63] [where P is the system for which Gödel proved his theorems.]

The situation is nevertheless *prima facie* a dilemma: one has to find an elementary, constructive proof (of a finitist statement) which cannot be carried out in S, but S is supposed to be adequate for elementary number theory. Thus, the metamathematical methods cannot be fixed once and for all, but must be suitably chosen for the specific theory. Further, the consistency of a part of mathematics cannot be settled with even "relative" finality, as a simple arithmetic statement, to wit, the consistency statement, can be formulated in S but not decided by S. This serious difficulty is already implied by the first incompleteness theorem. The very choice of the formal system S is problematic, as the completeness criterion is in principle not available in the essential cases.[64]

The conclusive and final mathematical solution to foundational problems Hilbert had hoped for cannot be obtained, the separation of mathematics from philosophy not sustained; these are the general consequences of Gödel's results. Yet if one gives up the dogmatic (metamathematical) restrictions and the radical (philosophical) aims, proof theory can be fruitfully developed. As early as 1932, Gödel and Gentzen independently gave a consistency proof for classical number theory Z. Their arguments showed that Z is reducible to intuitionistic arithmetic HA (in the sense of the introduction). As HA was seen to be correct from an intuitionistic viewpoint, a consistency proof had been given not by finitist but by intuitionistic means.[65] Bernays remarked, when looking back at this result in 1967:

> It thus became apparent that the "Finite Standpunkt" is not the only alternative to classical ways of reasoning and is not necessarily implied by the idea of proof theory. An enlarging of the methods of proof theory was therefore suggested: instead of a restriction to finitist methods of reasoning it was required only that the arguments be of a constructive character, allowing us to deal with more general forms of inference.[66]

Thus the question, Can we prove *the* system S for this part of mathematics to be consistent by finitist means? must be replaced by, Can this T, which is significant for this part of mathematics, be shown to be consistent by appropriate constructive means? Or, to paraphrase the new question, Is T reducible to a constructively justified theory? Hilbert and Bernays considered "ordinary analysis"[67] as the most significant T for which this modified goal should be pursued.

PART 2 – PROOF THEORY OF SUBSYSTEMS OF ANALYSIS

Two groups of (meta-) mathematical results will be discussed now. The results give (partial) answers to the questions I asked in my introductory remarks. Not surprisingly then, this part of my paper consists of two sections. In the first section I will describe formal theories for "ordinary" full classical analysis, some important subsystems, and a logical calculus in sequential form. In the second section I will outline some proof-theoretic investigations concerning elementary number-theory and impredicative subsystems of analysis. Here, I want to present matters in such a way that also the methods of proof are clearly indicated. As a matter of fact, I will restrict attention to one particularly lucid method due to G. Gentzen. It is my impression that it yields in the most direct and intelligible way partial solutions to Hilbert's reduction problem. For more information concerning this quite active field of research I have to refer to the literature; I mention in particular the survey papers of Kreisel and Feferman ([Kreisel, 1968], [Feferman, 1977]), and the Lecture Notes in Mathematics by Buchholz, Feferman, Pohlers, and Sieg.

2.1 Formalisms for (parts of) analysis

Hilbert gave a formalism for analysis in lectures during the early twenties. This formalism along with two equivalent formulations is presented in Supplement IV of *Grundlagen der Mathematik*, Band II. It is indicated there in rough sketches how to formally develop analysis and the theory of Cantor's second number class.[68] The main formalism is equivalent to second-order number theory with the full comprehension principle

$$\text{CA} \qquad (\exists Y)(\forall x)(x \in Y \leftrightarrow Fx)$$

and the axiom of choice in the form

$$\text{AC} \qquad (\forall x)(\exists Y)FxY \rightarrow (\exists Y)(\forall x)Fx(Y)_x,$$

where $y \in (Y)_x \leftrightarrow \langle y, x \rangle \in Y$, and $\langle, \rangle$ is a standard pairing function for natural numbers. In both CA and AC, F is an arbitrary second-order formula and may contain number- and set-parameters.[69]

Part of analysis can obviously be developed in subsystems of this theory. This is a trivial observation, as one uses in each specific proof only finitely many instances of the comprehension and choice principles. The interesting question is, whether significant portions can be captured in fixed, bounded parts. To discuss this point clearly we need a more detailed description of the formalisms.[70] All theories are formulated in the language $\mathscr{L}^2$ of second-order arithmetic; $\mathscr{L}^2$ is a two-sorted language with variables (and parameters) x, y, $z, \ldots (a, b, c, \ldots)$ ranging over natural numbers and X, Y, $Z, \ldots (A, B, C, \ldots)$ ranging over subsets of $\mathbb{N}$. As nonlogical symbols $\mathscr{L}^2$ contains 0, $'$, f_j for each $j \in \mathbb{N}$, $=$, $\in$, and $\perp$. $(f_j)_{j \in \mathbb{N}}$ is interpreted as an enumeration of the primitive recursive functions; they are assumed to be unary, except for the pairing function $\langle, \rangle$. $(\ )_0$ and $(\ )_1$ are the related projection functions. $\perp$ abbreviates $0 = 0'$. The numerical terms of $\mathscr{L}^2$ are built up as usual; s, t, $\ldots$ are syntactic variables for them. Formulas are obtained inductively from atomic formulas $s = t$ and $t \in A$ by closing under the propositional connectives $\rightarrow$, $\wedge$ and quantification over both sorts (and of both kinds). These logical symbols are chosen as the language is to serve for both classical and intuitionistic theories.[71] $F, G, \ldots$ are syntactic variables over formulas. The axioms for classical analysis (CA) are first of all those for zero and successor, pairing and projections; they include furthermore the defining equations for all one-place primitive recursive functions, the comprehension schema and the induction schema

$$F0 \wedge (\forall y)(Fy \rightarrow Fy') \rightarrow (\forall y) Fy$$

for all formulas F of $\mathscr{L}^2$. In the presence of CA, the induction schema can be replaced by the corresponding second-order axiom

$$(\forall X)[0 \in X \wedge (\forall y)\,(y \in X \rightarrow y' \in X) \rightarrow (\forall y)\, y \in X]$$

to yield an equivalent formal theory. However, as soon as the comprehension principle is restricted to a subclass of $\mathscr{L}^2$-formulas, the theory with the induction axiom is in general proof theoretically weaker than the theory with the schema. The base theory for our consideration is (PR-CA) $\upharpoonright$; i.e. second-order arithmetic with CA restricted to PR-formulas[72] and the induction axiom. That theory is conservative over primitive recursive arithmetic. The further second-order theories will extend this basic theory. They are denoted by the additional (set-existence) principle. $(\Pi^0_\infty\text{–CA}) \upharpoonright$ and $(\Pi^1_1\text{–CA}) \upharpoonright$ are (PR–CA) $\upharpoonright$ together with the comprehension principle for all arithmetic and Π^1_1-formulas, respectively; $(\Sigma^1_i\text{–AC}) \upharpoonright$ for $i = 1,\ 2$ is (PR–CA) $\upharpoonright$ extended by the instances of AC for all Σ^1_i-formulas.[73] Dropping " $\upharpoonright$ " from the name of a theory indicates that the full induction schema is available; for example, (PR–CA) is like (PR–CA) $\upharpoonright$ except that the induction axiom is replaced by the induction schema for all formulas of $\mathscr{L}^2$.

Hermann Weyl was the first to systematically develop analysis in a restricted framework. $\mathbb{N}$ is assumed, but subsets of the natural numbers have to be defined arithmetically; and that means now defined by a formula of $\mathscr{L}^2$ which contains only number-quantifiers. (Thus, impredicative definitions are avoided altogether.) Weyl's analysis, as presented in *Das Kontinuum*, can be formally carried out in $(\Pi^0_\infty\text{–CA})$. Feferman describes matters as follows in his lecture on 'Systems of Predicative Analysis':

The surprising result found by Weyl was that essentially the whole of analysis of continuous functions as contained, say, in the standard undergraduate course could be developed in this system [i.e., $(\Pi^0_\infty-CA)$]. We have, for example, for a continuous function on a closed interval of real numbers, existence of maximum and minimum, mean-value theorem, uniform continuity, existence of Riemann integral, and the Fundamental theorem of Calculus.[74]

The crucial mathematical point is that Dedekind's continuity principle can be proved in this theory when it is formulated only for sequences of arithmetically definable reals. The least-upper-bound principle, similarly restricted, can also be established. These principles suffice to prove appropriate versions of the Bolzano-Weierstrass and Heine-Borel theorems. (For this compare [Weyl, 1918], Kapitel II, §§ 4–5.)

Weyl's analysis can actually be carried out in $(\Pi^0_\infty-CA)\upharpoonright$; this fact is of substantial interest, as $(\Pi^0_\infty-CA)\upharpoonright$ is conservative over elementary number theory Z and remains so even if the choice principle for Σ^1_1–formulas is added.[75] Feferman [1977], Friedman [1976], and Takeuti [1978] have presented finite type theories which are also conservative over Z and which allow a more far-reaching and more convenient development. Essential here is that functions and sets are not interdefinable as in ordinary set theory. For answers to the obvious question, What cannot be done in such weak systems? I have to refer to the literature, e.g., [Feferman, 1977].

During the last decades, significant, stronger subsystems have been isolated; they are significant, because they allow the formalization of further substantial parts of mathematical practice, or admit natural nonmaximal models, or are based on foundational viewpoints more restrictive than, or alternative to, the set-theoretic one. A paradigm of such foundational research has been carried out under the heading of predicative analysis; see [Feferman, 1978]. In terms of Feferman and Schütte's precise characterization of "predicative analysis", the subsystem (Σ^1_1-AC), for example, is predicative. The theories $(\Pi^1_1-CA)\upharpoonright$ and (Σ^1_2-AC), in contrast, are impredicative. A constructive foundation for these theories will be given in section 2.2 below.

Quasi-empirical studies as those described above are, in my view, an indispensable component of work on the modified Hilbert program. The primary concern of proof theorists, however, has been the investigation of formal theories. In presenting a small part of such work, I will focus (as mentioned earlier) on Gentzen's method for analyzing the structure on derivations in formal and semiformal theories. It is based on special logical calculi, the so-called sequent calculi.[76] It is characteristic for them that they do not prove individual formulas (from assumptions), but rather pairs of finite sequences of formulas in the form

$$\Gamma \supset \Delta.$$

Γ and Δ stand for sequences $\langle F_1, \ldots, F_m \rangle$ and $\langle G_1, \ldots, G_n \rangle$, respectively. The *sequent* $\Gamma \supset \Delta$ may be interpreted as

$$(F_1 \wedge \ldots \wedge F_m) \rightarrow (G_1 \vee \ldots \vee G_n).$$

Let me describe a calculus L^2, appropriate for the language $\mathscr{L}^2$ of analysis. The calculus has as axioms the sequents

$$F \supset F,$$

where F is an atomic formula. There are two types of rules: STRUCTURAL and LOGICAL.

Structural Rules.

$$\frac{\Gamma \supset \Delta}{F, \Gamma \supset \Delta} \qquad \frac{\Gamma \supset \Delta}{\Gamma \supset \Delta, F}$$

$$\frac{F, F, \Gamma \supset \Delta}{F, \Gamma \supset \Delta} \qquad \frac{\Gamma \supset \Delta, F, F}{\Gamma \supset \Delta, F}$$

$$\frac{\Gamma \supset \Delta}{\Gamma^p \supset \Delta} \qquad \frac{\Gamma \supset \Delta}{\Gamma \supset \Delta^p}$$

Γ, F and F, Γ simply indicate the sequences obtained from Γ by pre-(suf-)fixing F. Γ^p is a permutation of the elements of Γ.

Logical Rules.

$$\frac{F_i, \Gamma \supset \Delta}{(F_1 \wedge F_2), \Gamma \supset \Delta} \quad i = 1 \text{ or } 2 \qquad \frac{\Gamma \supset \Delta, F \qquad \Gamma \supset \Delta, G}{\Gamma \supset \Delta, (F \wedge G)}$$

$$\frac{\Gamma \supset \Delta, F \qquad G, \Gamma \supset \Delta}{(F \rightarrow G), \Gamma \supset \Delta} \qquad \frac{\Gamma, F \supset G, \Delta}{\Gamma \supset \Delta, (F \rightarrow G)}$$

$$\frac{Ft, \Gamma \supset \Delta}{(\forall x)Fx, \Gamma \supset \Delta} \qquad *\frac{\Gamma \supset \Delta, Fa}{\Gamma \supset \Delta, (\forall x)Fx}$$

In the starred rule for the universal quantifier the parameter a must not occur in the lower sequent. There are clearly further rules, namely, dual ones for the existential number-quantifier and analogous ones for the set-quantifiers. Gentzen employed yet another rule to show that the sequent calculus is equivalent to a standard logical calculus of the Frege-Hilbert type, his SCHNITTREGEL or cut-rule:

$$\frac{\Gamma \supset \Delta, F \qquad F, \Gamma^1 \supset \Delta^1}{\Gamma, \Gamma^1 \supset \Delta, \Delta^1}$$

Remark. For the formalization of intuitionistic logic one requires that at most one formula occurs on the right-hand side (of $\supset$) in any sequent.

Gentzen proved a remarkable fact concerning the sequent calculus which he formulated as his HAUPTSATZ. It asserts that any derivation can be transformed into a cut-free derivation of the same endsequent.

THEOREM (Hauptsatz). If $\mathcal{D}$ is a derivation of $\Gamma \supset \Delta$, then we can find a cut-free derivation $\mathfrak{E}$ of $\Gamma \supset \Delta$.[77]

The proof provides a method to obtain finitistically the cut-free derivation $\mathfrak{E}$ from $\mathcal{D}$. The cut-elimination theorem is remarkable, as cut-free derivations exhibit a very special feature, namely, the subformula property: if $\mathcal{D}$ is a cut-free derivation of $\Gamma \supset \Delta$, then every formula occurring in $\mathcal{D}$ is a subformula of a formula in $\Gamma \supset \Delta$. Furthermore, some of the rules can be "inverted" in the cut-free part of the calculus, for example, the quantifier-rules. Let me state one such result.

INVERSION LEMMA.[78] Let $\mathcal{D}$ be a cut-free derivation of $\Gamma[(\forall x)Fx] \supset \Delta$, let $\Gamma(\Delta)$ contain only Π_1^0-(Σ_1^0-) formulas, and let Fa be quantifier-free; then we can find a sequence of terms $t_1, \ldots, t_n$ and a cut-free derivation $\mathfrak{E}$ of $\Gamma[Ft_1, \ldots, Ft_n] \supset \Delta$.

Let us quickly see how these results can be used to eliminate ideal statements from proofs of real ones, i.e., here quantifier-free statements without set-parameters. Z^- denotes the axioms for 0, $'$, f_j for all $j \in \mathbb{N}$, $\langle , \rangle$, $()_0$, $()_1$, i.e., Z^- is elementary number theory without induction. Assume F is a "real" statement in the above sense and can be proved from Z^-, i.e., there are finitely many axioms $A_1, \ldots, A_m$ and a derivation in L^2 of $\langle A_1, \ldots, A_m \rangle \supset F$. Then there is cut-free derivation $\mathcal{D}$ of that sequent, and indeed, by the inversion lemma, a cut-free derivation $\mathfrak{E}$ of

$$\langle A_1^*, \ldots, A_k^* \rangle \supset F,$$

where the A_j^* are instances of the Z^--axioms used in $\mathscr{D}$. Now we can observe that the endsequent is quantifier-free and (by the subformula property) *all* formulas in $\mathfrak{E}$ are sentential and purely arithmetic, i.e., real.[79]

2.2. *From Classical to Constructive Theories*

That is the direction work on the modified Hilbert program described in this section will take. Recall that in general the proof-theoretic equivalence of a classical theory T and a theory T^* of constructive character has to be shown, i.e., for each F in a class $\mathfrak{E}$ of formulas

(*) T proves F iff T^* proves F.

In the cases I will discuss, T^* is actually a subtheory of T and, consequently, one direction of (*) is trivial. Focusing on the other direction, we can first remark that it must be established constructively to be significant for Hilbert's program. Combining the constructivity requirements for the metamathematical argument and the theory T^*, the problem can be formulated more concisely as follows: we want to recognize the validity of proofs in T (for sentences F in $\mathfrak{E}$) from the standpoint of T^*. The technical problem is then to *establish in T^** all instances of the *partial reflection principle for T* and $\mathfrak{E}$

$$\mathbb{P}r_T(\ulcorner\mathscr{D}\urcorner, \ulcorner F\urcorner) \rightarrow F;$$

$\mathbb{P}r_T$ is the canonical representation of the proof predicate for T in T^*, $\ulcorner\mathscr{D}\urcorner$ and $\ulcorner F\urcorner$ are codes of a T-derivation $\mathscr{D}$ and a formula F, respectively.[80]

Let me look briefly at the paradigm of such a reduction, namely, Gödel and Gentzen's result concerning number theory. The proof of that result establishes finitistically that every negative arithmetic sentence provable in classical number theory can already be proved intuitionistically.

Remark. This proof can be extended to theories of iterated inductive definitions $\mathrm{ID}_{<\nu}$ and full classical analysis (CA); they, too, are conservative over their formally intuitionistic versions with respect to all negative arithmetic sentences.[81] In the rest of this section I will explain, how the $\mathrm{ID}_{<\nu}$ can be reduced to a constructive theory. That solves the reduction problem for the impredicative theories $(\Pi_1^1-CA)\upharpoonright$ and (Σ_2^1-AC), as they are proof-theoretically equivalent to $ID_{<\omega}$ and $ID_{<\epsilon_0}$, respectively, by work of Feferman [1970] and Friedman [1970]. However, let me make one step at a time!

How can it be seen that HA proves all instances of the partial reflection principle for Z and all negative arithmetic sentences? The first additional step is a direct formalization of the metamathematical argument *in HA*; that is easily done and gives in HA

$$\mathbb{P}\,r_Z(\ulcorner\mathscr{D}\urcorner, \ulcorner F\urcorner) \to \mathbb{P}\,r_{HA}(\sigma(\ulcorner\mathscr{D}\urcorner), \ulcorner F\urcorner);$$

σ is a primitive recursive function transforming classical into intuitionistic proofs. In the second step one makes use of the following facts: (i) $\sigma(\ulcorner\mathscr{D}\urcorner)$ contains only a finite number of formulas and their complexity is consequently bounded by a fixed n; (ii) for formulas of bounded complexity one can give in HA a partial truth-definition T_n. The adequacy of T_n, i.e.,

$$T_n(\ulcorner F\urcorner) \leftrightarrow F,$$

can be established in HA for formulas of complexity $\leq n$, and so can the soundness of derivations containing only such formulas, i.e.,

$$\mathbb{P}\,r_{HA}(a, \ulcorner F\urcorner) \to T_n(\ulcorner F\urcorner).$$

Detaching in the obvious way we obtain the partial reflection principle.[82]

The argument I just sketched can be carried out for any equivalent formalization of Z and HA. Its first step can be taken in primitive recursive arithmetic PRA, the quantifier-free part of Z or, equivalently, HA. In the second step HA is required for recognizing the truth of axioms (in particular, instances of induction), and simply to formulate the partial truth-definition for quantified formulas. This is in general so, even if the end-formula of a derivation $\mathscr{D}$ is quantifier-free: $\mathscr{D}$ may contain quantified formulas. Let me show now in a simple example, how the sequent calculus and the Hauptsatz are used to eliminate such "extraneous elements" from proofs. For this purpose I return to and continue the investigation of Z^- at the end of section 2.1 above. The considerations presented there, including the proof of the Hauptsatz, can be formalized in PRA to yield

$$\mathbb{P}\,r_Z(\ulcorner\mathscr{D}\urcorner, \ulcorner F\urcorner) \to \mathbb{P}\,r(\ulcorner\mathscr{E}\urcorner, \ulcorner\langle A_1^*, \ldots, A_k^*\rangle \supset F\urcorner).$$

Due to the subformula property of cut-free derivations, the (purely sentential) complexity of formulas in $\mathscr{E}$ is bounded by that of the formulas in the endsequent. A partial truth-definition for quantifier-free formulas of bounded complexity can be defined and shown to be adequate in PRA. Defining

$$T_n(\ulcorner \Delta \supset G \urcorner) \leftrightarrow (T_n(\ulcorner F_1 \urcorner) \wedge \cdots \wedge T_n(\ulcorner F_m \urcorner) \rightarrow T_n(\ulcorner G \urcorner))$$

where Δ is $\langle F_1, \ldots, F_m \rangle$ and all formulas in $\Delta \supset G$ are of complexity $\leqslant n$, one shows

$$\mathbf{P}r(a, \ulcorner \Delta \supset G \urcorner) \rightarrow T_n(\ulcorner \Delta \supset G \urcorner).$$

So we can immediately infer

$$T_n(\ulcorner A_1^* \urcorner) \wedge \cdots \wedge T_n(\ulcorner A_k^* \urcorner) \rightarrow T_n(\ulcorner F \urcorner),$$

and then F, making use of T_n's adequacy and the provability in PRA of the A_j^*. In summary, we have obtained in PRA the partial reflection principle for Z^- and quantifier-free formulas F

$$\mathbb{P}r_{Z^-}(\ulcorner \mathscr{D} \urcorner, \ulcorner F \urcorner) \rightarrow F.$$

The considerations concerning (and involving) and partial truth definition T_n are so obvious for the simple theory Z^-, that Gentzen remarked after having described the transformation of $\mathscr{D}$ into a cut-free derivation $\mathscr{E}$:

That a contradiction cannot be inferred from such statements [i.e., quantifier-free ones of Z^-] by means of sentential logic is almost self-evident (selbstverständlich); and a proof of that would hardly be more than a formal description (Umschreibung) of an intuitively clear fact (Sachverhalt).[83]

Nevertheless, by making this very last step explicit one has the schema of an argument which can be adapted to more complicated situations. Indeed, it can be extended easily to the theory Z_0, i.e., Z^- together with the induction schema for quantifier-free formulas. The general induction schema, however, is a genuine block to a straightforward extension. Gentzen overcame the problem in his 1936 and 1938 consistency proofs for Z. He used in those proofs in addition to finitist principles formulated in PRA the transfinite induction and recursion principles up to the first ϵ-number ϵ_0.[84]

The role of transfinite ordinals in Gentzen's proofs was at first difficult to understand and to assess. Today, after Lorenzen's, Schütte's, and Tait's work, their role (in a version of the 1938 proof) is perfectly clear:

they serve as a natural measure of the length of formulas and derivations in an infinitary propositional calculus! Let me explain this claim by outlining the argument as it is now standardly presented. The infinitary calculus PL is obtained as an extension of the propositional and purely number-theoretic part of L^2 by admitting conjunctions and disjunctions of infinite sequences of formulas: if $(A_n)_{n \in \mathbb{N}}$ is a sequence of formulas, then $\Pi_n A_n$ and $\Sigma_n A_n$ are also formulas. The rules for Π are the following:

$$\frac{\Gamma \supset \Delta, A_n}{\Gamma \supset \Delta, \Pi_n A_n} \qquad \text{for each } n \in \mathbb{N},$$

$$\frac{A_n, \Gamma \supset \Delta}{\Pi_n A_n, \Gamma \supset \Delta} \qquad \text{for some } n \in \mathbb{N}.$$

The rules for Σ are dual. The axioms of PL are the sequents $\Delta \supset A$ and $A \supset \Delta$, where Δ is the empty sequence and A is a true and a false atomic sentence, respectively.

The sentences of the language of Z can be translated into formulas of PL, the infinitary operators replacing quantifiers. For example, if Fa is quantifier-free, then $(\forall x)Fx$ is translated as $\Pi_n Fn$.[85] Z-derivations (of sentences) can be transformed into "purely logical" PL-derivations, as all Z-axioms are provable in PL. That holds in particular for the instances of the induction principle and is seen most easily when the induction rule

$$\frac{F0 \qquad (\forall x)(Fx \to Fx')}{(\forall x)Fx}$$

is used.[86] For assume that $\mathscr{D}_1$ and $\mathscr{D}_2$ are Z-derivations of $F0$ and $(\forall x)(Fx \to Fx')$, respectively; then one has (in the inductive argument on the length of derivations) PL-derivations $\mathscr{E}_1$ and $\mathscr{E}_2$ for $F0$ and $\Pi_n (Fn \to Fn')$.[87] $\mathscr{E}_1$ and $\mathscr{E}_2$ permit us to construct an infinite PL-proof of $\Pi_n Fn$, which is roughly indicated by

$$\mathscr{E}_1 \left\{ \begin{array}{c} \downarrow \\ F0 \end{array} \right. \quad \frac{\begin{array}{cc} \downarrow & \Pi_n(Fn \to Fn') \Big\} \mathscr{E}_2 \\ F0 & F0 \to F1 \end{array}}{F1} \quad \frac{\begin{array}{c} \downarrow \\ F0 \end{array} \quad \dfrac{\begin{array}{c} \downarrow \\ \Pi_n(Fn \to Fn') \\ \hline F0 \to F1 \end{array}}{F1} \quad \dfrac{\begin{array}{c} \downarrow \\ \Pi_n(Fn \to Fn') \\ F1 \to F2 \end{array}}{}}{F2} \cdots$$

$$\rule{8cm}{0.4pt}$$
$$\Pi_n Fn$$

Measuring the length of formulas and derivations of PL with ordinals of the second number class, it is straightforward to prove the cut-elimination theorem for PL by transfinite induction. Consequently, Z-derivations can be transformed into cut-free PL-derivations. If the endformula F of a Z-derivation is Π_1^0, then the associated cut-free PL-derivation contains only subformulas of (the translation of) F. For such formulas one can give an adequate partial truth-definition even in PRA, as we have seen; and the truth of F can be established by induction on the length of its cut-free PL-derivation. The argument schema is the same as for the treatment of Z^-; but are the considerations here not highly nonconstructive? Due to the fact that Z is a formal theory, it is sufficient to take into account an effectively described part of PL, all of whose derivations are less than $\in_0$. The argument sketched above can actually be formalized in PRA + TI(β), $\beta < \in_0$.[88] In summary then, each instance of the partial reflection principle for Z and quantifier-free formulas F

$$\mathbb{P}\, r_z(\ulcorner \mathcal{D} \urcorner, \ulcorner F \urcorner) \to F$$

can be proved in PRA + TI(β) for some $\beta < \in_0$.[89]

The investigation of formal theories via infinitary sequent calculi proved extremely fruitful for predicative analysis. It turned out that the approach can be extended to impredicative subsystems of analysis, seemingly in the most understandable and informative way, through an investigation of proof-theoretically equivalent theories of transfinitely iterated inductive definitions.[90] The idea of an inductive definition is familiar from mathematics and mathematical logic; for example, the formulas of a formal language and the theorems of a formal theory are determined inductively. The one feature which gives inductively defined classes (i.d. classes) their special character and accounts in part for their foundational interest is captured by saying that they are generated according to rules or that they are obtained by iterated application of a definable operator.[91] Let me mention one i.d. class which is not given by an "elementary" inductive definition (as those above), namely, the class $\mathcal{O}_0$ of constructive ordinals. That class is given by two inductive clauses: (i) 0 is in $\mathcal{O}_0$; (ii) if a is (the Gödel-number of) a recursive function enumerating elements of $\mathcal{O}_0$, then a is in $\mathcal{O}_0$. The elements of $\mathcal{O}_0$ are thus generated by joining recursively given sequences of previously generated elements of $\mathcal{O}_0$ and can be pictured as infinite, well-founded trees. As this construction-tree is uniquely determined for each constructive ordinal a, it can be viewed as a canonical proof (built by using only rules (i) and (ii)) showing that a is in $\mathcal{O}_0$. Locally, the structure of a tree is as follows:

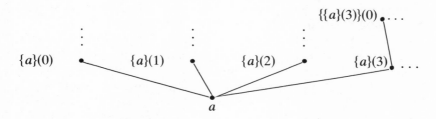

By iterating the definition of $\mathcal{O}_0$ along a given recursive well-ordering $<$ of the natural numbers one obtains the higher constructive number classes $\mathcal{O}_\nu$.[92] Without stating the inductive clauses let me just say that the elements of $\mathcal{O}_\nu$ can be pictured as infinite, well-founded trees, which branch not only over $\mathbb{N}$ but also over number classes $\mathcal{O}_\mu$, $\mu < \nu$.

The theory for $\mathcal{O}_0$, denoted by $\mathrm{ID}_1(\mathcal{O})$, is an extension of elementary number theory. Its language is that of Z expanded by the unary-predicate symbol $\mathcal{O}_0$, its axioms those of Z (with induction for all formulas in the expanded language) together with two principles for $\mathcal{O}_0$. For their formulation let me note first, that the clauses (i) and (ii) above can be given by a formula $\mathfrak{A}$ which is arithmetic and positive in $\mathcal{O}_0$.[93] The first principle is a closure principle and expresses that $\mathfrak{A}$ leads from elements of $\mathcal{O}_0$ to elements in $\mathcal{O}_0$:

$(\mathcal{O}.1)$ $\mathfrak{A}(\mathcal{O}_0) \subseteq \mathcal{O}_0$;

the second principle is a minimality principle and expresses that one can give arguments by induction on $\mathcal{O}_0$:

$(\mathcal{O}.2)$ $\mathfrak{A}(F) \subseteq F \rightarrow \mathcal{O}_0 \subseteq F.$

(In other words, if the class given by a formula F is closed under $\mathfrak{A}$, then $\mathcal{O}_0$ is contained in that class.) The formal principles for the higher number classes $\mathcal{O}_\mu$, $\mu < \nu$, are formulated in $\mathrm{ID}_\nu(\mathcal{O})$ in analogy to those for $\mathcal{O}_0$.

$(\mathcal{O}.1)_\nu$ $(\forall y < \nu)\mathfrak{A}_y(\mathcal{O}_y) \subseteq \mathcal{O}_y$;

$(\mathcal{O}.2)_\nu$ $(\forall y < \nu)\,(\mathfrak{A}_y(F) \subseteq F \rightarrow \mathcal{O}_y \subseteq F)$

The theory $ID_<(\mathcal{O})$ is just the union of the $ID_\nu(\mathcal{O})$; it is also denoted by $ID_{<\alpha}$, if the classical ordinal associated with $<$ is α. In the general theories of iterated inductive definitions ID_ν and $ID_<$ one has these principles for all i.d. classes, which are determined by arbitrary positive defining clauses iterated along $<$. The intuitionistic versions of these theories are denoted by $ID^i_\nu(\mathcal{O})$, $ID^i_<(\mathcal{O})$, ID^i_ν, and $ID^i_<$.

Remark. These theories have recently been analyzed from a proof-theoretic point of view by Buchholz, Pohlers, and myself. The various results we obtained will be published in the LN-volume already metioned. In the following I focus on a reductive result in my dissertation [Sieg, 1977].[94]

The reductive result to be described now is this: assuming that α is a recursive ordinal of limit characteristic, the classical theory of arbitrary inductive definitions $ID_{<\alpha}$ is conservative over the intuitionistic theory of higher constructive number classes $ID^i_{<\alpha}(\mathcal{O})$ for all negative arithmetic and Π^0_2-sentences. The crucial step in the reduction is a Gentzen-type analysis of ID_ν, $\nu < \alpha$, (which can also be carried out for ID^i_ν) showing that $ID_{<\alpha}$ and $ID_{<\alpha}(\mathcal{O})$ prove exactly the same arithmetic sentences. The schema of the argument is that of the proof-theoretic analysis of Z^- and Z.

For the investigation of number theory the infinitary propositional calculus PL was used. Even stronger infinitary calculi PL_ν are considered here; conjunctions and disjunctions are taken over $\mathbb{N}$ as well as over all number classes $\mathcal{O}_\mu, \mu < \nu$. The language of PL_ν allows us to define explicitly all i.d. classes given by a positive inductive definition iterated at most ν times. This is the basis for a translation of the language of ID_ν into that of PL_ν, associating with each formula F in the former a formula F^+ in the latter. A derivation $\mathcal{D}$ of F in ID_ν can be turned into a PL_ν-*derivation of* $CP \supset F^+$, where CP is the translation of the closure principle for the i.d. class mentioned in $\mathcal{D}$.[95] The natural metatheory for these considerations is $ID_{\nu+1}(\mathcal{O})$, as the syntactic objects of PL_ν just are trees in $\mathcal{O}_\nu$. In this theory one can prove without using a special system of ordinal notations the cut-elimination theorem for PL_ν. Thus, the PL_ν-derivation of $CP \supset F^+$ can be assumed to be cut-free. From such derivations one can eliminate CP, if F is arithmetic. (The proof of this elimination lemma is a little delicate, but can also be carried out in $ID_{\nu+1}(\mathcal{O})$.) Using again a partial truth-definition and the soundness of cut-free proofs, the truth of F^+ can be established. So we obtain for ID_ν and any arithmetic F the partial reflection principle in $ID_{\nu+1}(\mathcal{O})$, analogously that for ID_ν^i in $ID_{\nu+1}^i(\mathcal{O})$. We know now in particular that the partial reflection principle for $ID_{<\alpha}^i$ and all arithmetic formulas

$$(*) \qquad \mathbb{P}r_{ID_{<\alpha}^i}(\ulcorner\mathscr{D}\urcorner, \ulcorner F\urcorner) \to F$$

can be proved in $ID_{<\alpha}^i(\mathcal{O})$.

The argument for the reduction of $(\Pi_1^1\text{-CA}){\upharpoonright}$ and $(\Sigma_2^1\text{-AC})$ to constructive theories can be quickly concluded now. It was mentioned in the *Remark* at the beginning of this subsection that $(\Pi_1^1\text{-CA}){\upharpoonright}$ and $(\Sigma_2^1\text{-AC})$ are proof-theoretically equivalent to $ID_{<\omega}$ and $ID_{<\epsilon_0}$ respectively; these latter theories can be reduced to their intuitionistic versions by an extension of the double-negation translation. As the steps involved are finitist and can be taken in PRA, one certainly has in $ID_{<\epsilon_0}^i(\mathcal{O})$ for negative arithmetic and Π_2^0-sentences F

$$(**) \qquad \mathbb{P}r_{(\Sigma_2^1\text{-AC})}(\ulcorner\mathscr{D}\urcorner, \ulcorner F\urcorner) \to \mathbb{P}r_{ID_{<\epsilon_0}^i}(\rho(\ulcorner\mathscr{D}\urcorner), \ulcorner F\urcorner);^{96}$$

combining (*) with (**) yields the desired partial reflection principle for $(\Sigma_2^1\text{-AC})$ in $ID_{<\epsilon_0}^i(\mathcal{O})$ for the above class of sentences;

$$\mathbb{P}r_{(\Sigma_2^1\text{-AC})}(\ulcorner\mathscr{D}\urcorner, \ulcorner F\urcorner) \to F.$$

Similarly, one can prove the partial reflection principle for $(\Pi_1^1\text{-CA}){\upharpoonright}$ in $ID_{<\omega}^i(\mathcal{O})$.[97]

Detailed axiomatic investigations of the type reported in section 2.1 above are of obvious mathematical and philosophical interest. But what has been achieved by the reductions of impredicative theories for analysis to theories for ordinals? Any answer which focuses on their foundational significance will certainly have to argue for the constructive character of the theories for ordinals, and that can be described in part as follows. The theories are based on intuitionistic logic; the objects in their intended models are obtained by construction; the definition– and proof-principles which are admitted in the theories follow that construction. The objects, i.e., the constructive ordinals, are furthermore of a very special character. They reflect their buildup according to the generating clauses of their definition in a direct and locally effective way. Viewing the clauses as inference rules, the constructive ordinals are infinitary derivations and show that they fall under their definition. All of this indicates that the theories for ordinals are constructively justified and thus provide a constructive foundation for the classical theories which are reducible to them.

This answer is complemented and its claim (perhaps) supported by technical results of the following form: there is a recursive well-ordering $<$ and a theory T^* for constructive analysis, such that $\mathrm{ID}^i_<(\mathcal{O})$ and T^* are proof-theoretically equivalent. Some results of this form have been obtained by Troelstra and by Feferman and Sieg.[98] The conceptually important point, which is technically exploited for their proof, is that sufficiently strong theories for constructive analysis (should) contain principles allowing the construction of ordinals. (It is the parenthetical "should" which accounts for the "perhaps" at the beginning of this paragraph.) The reductive results make it possible now to compare the proof-theoretic strength of theories for classical and constructive analysis. That can lead to highly important insights into the proof-theoretic structure of parts of mathematics.[99]

So much for the more immediate foundational significance of theories for ordinals (in connection with the modified Hilbert program) and the almost direct application of the reductive results. From a broader perspective, I see these investigations as part of an attempt to take the concept of iteration or inductive definition as basic for analyzing that section of mathematics which lends itself to an arithmetic, constructive treatment.[100] It would be a step forward if a more general notion of i.d. class could be found, permitting distinctions which are mathematically intelligible and significant.[101] Work in two areas might provide complementary experience helpful for finding such a notion and suitable distinctions: (1) the proof theory of subsystems of ZF, and (2) the detailed study of the second constructive number class. Jäger and Pohlers have started work on (1); for (2) I have a concrete problem in mind. When discussing Spector's consistency proof for full classical analysis and the possibility of obtaining satisfactory consistency proofs for parts of analysis, Gödel noted:

Perhaps the most promising extension of the system T for the Dialectica-interpretation of HA is that obtained by introducing higher type computable functions for constructive ordinals.[102]

It seems to me that for such an extension one must find new constructive principles for ordinals of the *second number class*, which can be used to give Dialectica-interpretations of intuitionistic theories of ordinals $\mathrm{ID}^i_<(\mathcal{O})$ for various well-orderings $<$.

CONCLUDING REMARKS

Classical mathematical analysis can be developed in a conservative extension of elementary number theory; some impredicative subsystems of analysis do have a constructive foundation. These are, very informally stated, the main facts which are established by the work described in Part 2. They throw an ironic light on the dispute between Brouwer and Hilbert during the twenties. Keep in mind that both men intended to provide a constructive basis for mathematics, with the crucial difference that Brouwer wanted to develop mathematics exclusively by intuitionistic principles "in unerschütterlicher Sicherkeit", whereas Hilbert strove for a justification of (the use of) classical mathematics through finitist means.[103] They agreed, however, on one fundamental point, namely, that some assumptions of classical mathematics transcend what is given by elementary intuition. The rejection of such assumptions would lead, they both expected, to severe restrictions of classical mathematics and, indeed, of analysis. That conclusion is certainly not supported by section 2.1 above.[104]

The second fact mentioned above is not so much related to (the effect of) the rejection of classical assumptions as to the strength of the admitted constructive principles. The incompleteness theorems showed that finitist mathematics cannot support Hilbert's program; Gödel and Gentzen's consistency proof for classical number theory, on the other hand, made it clear that intuitionistic principles go beyond finitist ones. It is most doubtful that they are all justified by what Brouwer called the "fundamental phenomenon of mathematical thinking, the intuition of the bare two-oneness".[105] Indeed, by section 2.2 we know that principles which are intuitionistically acceptable permit the reduction of impredicative classical theories – theories so problematic from a constructive point of view.[106]

Bernays pointed out that the intuitionistic understanding of conditional statements is not based on elementary evidence when their antecedents are universal statements or conditionals. The reason, Bernays argues, is that intuitionistic methods of proof are not fixed;[107] they are most decidedly not captured in a formal theory. As a matter of fact, they included infinitary proofs which were admitted and investigated by Brouwer. Such proofs are mathematically treated as constructive, well-founded trees or ordinals. With reference to them Brouwer wrote:

These *mental* [gedankliche] mathematical proofs that in general contain infinitely many terms must not be confused with their linguistic accompaniments, which are finite and necessarily inadequate, hence do not belong to mathematics.[108]

He continued that this remark contains his "main argument against the claims of Hilbert's metamathematics".[109]

Zermelo, from a completely different foundational viewpoint, also asserted that finite linguistic means are inadequate for capturing the essence of mathematics and of mathematical proofs:

... "combinations of symbols" are *not*... the true subject matter of mathematics, but *conceptual-ideal relations* between the elements of a conceptually posited *infinite manifold*; and our systems of symbols are in this always just *imperfect* ... *auxiliary means* of our *finite* mind, to conquer at least in stepwise approximation the infinite, which we cannot *directly* and *intuitively* "survey" or grasp.[110]

To overcome finitist restrictions he proposed an infinitary logic and mathematical theories which ensure that, for a given infinite domain of elements and relations between them, truth and provability coincide. Proofs are infinite well-founded trees and represent, very roughly speaking, classical truth-conditions. The epistemologically problematic aspect of such infinitary proofs is quite clearly seen by Zermelo:

Such a "proof" contains most often *infinitely many* intermediate sentences, and it is not clear yet, in how far and by which auxilary means it [such a proof] can be made intelligible also to our *finite* mind. Basically *any* mathematical proof, e.g. the inference principle [Schlußverfahren] of "complete induction", is thoroughly "infinitistic", and yet we are able to comprehend it. There seem to be no fixed limits of intelligibility here.[111]

The effect of infinitary derivations can be partially obtained in a "finitist" manner. After all, some infinitistic principles (e.g., induction principles, choice principles) can be formulated, even if only inadequately, in formal languages and can be employed in finite derivations for the recognition of truths.[112] For precisely this purpose Gödel suggested the use of stronger and stronger axioms of infinity, as he saw the true reason for the incompleteness of formal systems in the fact that "the formation of ever higher types can be continued into the transfinite ..., while in any formal system at most denumerably many of them are available."[113] When investigating subsystems of analysis in section 2.2 above , we proceeded in exactly the opposite direction: finite derivations, in which infinitistic principles are used, were expanded into infinitary proofs. Such proofs were transformed into cut-free ones and then recognized as correct (intuitionistic) truth-conditions for their endformulas. These considerations were carried out by constructive, but certainly nonelementary principles.[114]

Gödel remarked that giving a constructive consistency proof for classical mathematics means

to replace its axioms [those of classical mathematics] about abstract entities of an objective Platonic realm by insights about the given operations of our mind.[115]

This poignant and schematic formulation assigns to constructive consistency proofs the task of relating two essential aspects of our mathematical experience:

– the impression that mathematics deals with abstract objects arranged in structures which are independent of us;
– the conviction that principles for some structures are a priori and more immediately evident, because the buildup of their elements corresponds so intimately to operations of our mind that we even say the objects are created by us.[116]

If constructive consistency proofs are viewed in this light, they can still be said to provide epistemological reductions, though certainly not in the strong justificatory sense dear to Hilbert. What seems to me to be essential for gaining a general perspective is the fact that they are (usually the most difficult) part of establishing equivalences between theories which are based on prima facie radically opposed, irreconcilable foundational viewpoints. The very fact that such equivalences can be given for substantive theories, I believe, undermines traditional positions and encourages an open view for mathematical facts and an unforced attitude in foundational work.[117]

The investigations on the foundations of mathematics are being pursued vigorously. Several basic questions remain open, and we do not know what more we are destined to discover in this area. In any event, these investigations arouse in their changing aspects our curiosity – a sentiment which is brought forth only to a lesser degree by the classical areas of mathematics, which have already achieved a greater perfection.[118]

NOTES

* This paper was completed in June 1981; some minor changes were made in August 1982.
[1] [Bernays, 1930], p. 17.
[2] With one exception: Frege contemplated in 1924/25 to base arithmetic on geometry.
[3] N. H. Abel, *Memorial*, Kristiana, 1902, p. 23
[4] [Gauss, 1880], p. 497.
[5] [Dedekind, 1932/1888], p. 338. (I refer always to the third volume of Dedekind's

collected papers; "1888" after the slash indicates that the quotation is taken from "Was sind und was sollen die Zahlen?" and "1872" that the quotation is from "Stetigkeit und irrationale Zahlen".)

[6] [Dedekind, 1932/1872], p. 322.

[7] In his essay Dedekind speaks of "creating" new numbers; in a letter to Lipschitz, [Dedekind, 1932] p. 471, he emphasizes that one could "identify" the real numbers with cuts.

[8] [Dedekind, 1932], p. 473. In another letter to Lipschitz Dedekind points out the crucial difference between Euclid's treatment of irrational magnitudes and his own. "Euclid can apply his definition of equal proportion to magnitudes, which come up in his system, i.e. whose existence is evident [ersichtlich] for good reasons, and that is quite sufficient for Euclid." However, Dedekind – in contrast to Euclid – wants to base arithmetic on the concept of magnitude, and thus it is crucial to know "from the beginning how complete (continuous) the domain of magnitudes is, because nothing is more dangerous in mathematics, than to *assume* without sufficient proof existences . . . ." [Dedekind, 1932], p. 477.

[9] This theorem is not mentioned in Dedekind, but it motivated Bolzano to search for a purely analytic proof; and such a radical modern constructivist as Bishop finds it "intuitively appealing", but unprovable in his theory. The second theorem which is discussed by Dedekind is Cauchy's convergence criterion for a function from R to R.

[10] [Dedekind, 1932/1888], p. 339.

[11] [Dedekind, 1932/1888], p. 335. Frege's views are clearly parallel; he speaks of "defining the real numbers purely arithmetically or logically". (*Grundgesetze der Arithmetik*, II. Band, 1903, p. 162.) Cp. also footnote 1), ibid. p. 155.

[12] .[Kronecker, 1887] in [Kronecker, *Werke* III], p. 260. As to Kronecker's understanding of "arithmetic" see the end of section 1.2. Here it is to be noted, that the general notion of irrational number was excluded in principle from arithmetic proper. Kronecker's paper was actually part of the reason for Dedekind's decision to publish his thoughts on the matter. It is also informative to hear how Dedekind's essay was received in Berlin." In Berlin, Dedekind's essay "Was sind und was sollen die Zahlen?", which had just been published, was talked about in all mathematical circles by young and old – mostly in a hostile manner." Hilbert, who had traveled in 1888 to various German universities, reported this impression in [1931, p. 487].

[13] [Poincaré, 1902] p. 120 and p. 122.

[14] [Dedekind, 1932/1888] p. 357.

[15] Cantor's letter has been translated and can be found in [van Heijenoort, 1967].

[16] [Hilbert, 1900], reprinted also in *Grundlagen der Geometrie*. The paper is dated "Göttingen, den 12. Oktober 1899".

[17] At the end of the paper, Hilbert claims that if matters are viewed his way then the reservations concerning the existence of the totality of real numbers lose all their justification.

[18] [Hilbert, 1922/1900], p. 242. Note that "consistency proof" corresponds to "Beweis der Widerspruchslosigkeit", "consistent set" to "konsistente Menge". As to my claim that Hilbert thought of Dedekind's and Weierstrass's method, see [Bernays, 1967], p. 500.

[19] [Hilbert, 1902], pp. 73–74. These views seem to me to be quite obscure; in particular, does the axiom system have to satisfy further conditions, apart from consistency? In [Hilbert, 1922/1900], p. 242, "endlich", "abgeschlossen", and syntactic completeness (?)

are mentioned; are these properties of the axiomatization thought to be important?

[20] Cp. Zermelo's footnote 9 on p. 191 of [van Heijenoort, 1967] and [Bernays, 1935], p. 199.

[21] [Zermelo, 1908] and [Russell, 1908].

[22] So Zermelo in a report to the Emergency Society of German Science, reprinted in [Moore, 1980].

[23] [Zermelo, 1908], p. 200.

[24] Ibid., p. 201.

[25] Ibid., p. 200.

[26] For penetrating discussions see [Gödel, 1944] and [Gandy, 1973]. In the latter reference in particular the section "Philosophical Framework".

[27] [Russell, 1908], p. 167.

[28] Ibid., p. 168.

[29] [Whitehead, Russell, 1910], p.

[30] See for example Chapter XIX of Russell's *Principles of Mathematics*.

[31] In [1930], p. 31, Zermelo considers the axiom of choice as a "general logical principle".

[32] The letters are reprinted in [Borel, 1914], pp. 150–160.

[33] Ibid., p. 153.

[34] See Kronecker's "Über einige Anwendungen der Modulsysteme auf elementare algebraische Fragen", *Journal für Mathematik* **99**, 1886, pp. 334–336, in particular footnote * on p. 336.

[35] One has only to think of Cantor's complaints, Weierstrass's remarks, and Hilbert's observations.

[36] The main mathematical work, however, is devoted to an effective analysis of the real roots of algebraic equations.

[37] And that holds no matter whether they are created by God (as remarked by Kronecker in a talk at the Berliner Naturforscher-Versammlung in 1886 [Weber, 1893], p. 19) or whether the concept of number is being developed in philosophy *before* it is investigated by mathematicians (as suggested by Kronecker at the beginning of his essay).

[38] [Kronecker, 1887], p. 253; cp. also the remarks on p. 274.

[39] Ibid., p. 272.

[40] [Kronecker, 1901], p. VI.

[41] One should, however, be aware of the quite different philosophical visions.

[42] Due to the *intuitionistic* continuity principle.

[43] Choice sequences were introduced to obtain a more adequate theory of the continuum.

[44] [Reid, 1970], p. 173.

[45] [Hilbert, 1904], p. 258.

[46] [Hilbert, 1904], pp. 252–253, 257–258.

[47] In this proposal two different tasks are involved: (i) an axiomatization of set-theoretic principles, and (ii) a strictly formal development of the subject.

[48] [Hilbert, 1904], pp. 251–252 and V on p. 257.

[49] [Hilbert, 1935] p. 160. This is Hilbert's "Neubegründung der Mathematik. Erste Mitteilung".

[50] [Bernays, 1967], p. 500.

[51] With regard to my interpretative claim, see [Bernays, 1922A], p. 15. The conflicting

doctrines may be said to "correspond" to opposing tendencies in Hilbert's own thinking about mathematics. "On one side, he [Hilbert] was convinced of the soundness of existing mathematics; on the other hand, he had – philosophically – a strong scepticism. "See Bernays in [Reid, 1970], p. 173.

[52] The existence of mathematical structures is no longer guaranteed by a consistency proof – even not programmatically. They play, however, a role in mathematical thinking. This is a part of Hilbert's and, especially, Bernays's view which I find extremely fascinating – AND puzzling. See for example the pre-Gödel paper [Bernays, 1930], pp. 54–55 and [Bernays 1950] which is devoted to precisely this problem.

[53] As to Kronecker, see [Hilbert, 1931], p. 487; as to Brouwer, see [Bernays, 1930], pp. 41–42 and fn. 9 on p. 42.

[54] [Bernays, 1922A], p. 12.

[55] Cp. with [Kreisel, 1968], pp. 322–323.

[56] [Hilbert, Bernays, 1934], p. 42 and p. 44.

[57] This equivalence is easily established under very general conditions on S; for details see Smorynski's, The incompleteness theorems, Handbook of Mathematical Logic, J. Barwise (ed.) North-Holland, Amsterdam, 1977, p. 846, Theorem 4.1.4.

[58] E.g., in Hilbert's Grundlagen der Mathematik, translated in [van Heijenoort, 1967], p. 474.

[59] Ibid., p. 465 in translation.

[60] [Hilbert, 1928], p. 12 and [Bernays, 1930], p. 58; Hilbert's paper was delivered at the International Congress of Mathematicians in Bologna.

[61] [Bernays, 1930], p. 59. It is important to notice here that Hilbert and Bernays did not think that any formal theory would be a final, nonextendible framework for mathematics; it would be extendible by new concepts! Nevertheless, Bernays argues that a formal theory may be such that an extension by new concepts does not lead to new results in the original (language of the) theory. This condition, he continues, is certainly satisfied if the theory is deductively closed, i.e., each sentence, which can be formulated in the theory, is either provable or refutable. It is precisely syntactic completeness in this sense which is excluded by Gödel's first incompleteness theorem.

[62] [Bernays, 1935], p. 211.

[63] [Gödel, 1931], p. 37.

[64] Gödel emphasized this point in his 1931 correspondence with Zermelo, published in Gratten-Guinness, 'In memoriam K. Gödel', Hist. Math., 6 (1979), 294–304. Cp. also note 61.

[65] This was incidentally quite revealing, as it had been assumed in the Hilbert school that intuitionist and finitist reasoning were identical. Cp. note 53.

[66] [Bernays, 1967], p. 502.

[67] Their formalism of (ordinary) analysis is described at the beginning of section 2.1

[68] For someone familiar with the development of classical analysis in ZFC it is easy to see that in such a presentation one appeals only to the first three levels of the cumulative hierarchy, considering the natural numbers as urelements. These levels are prima facie needed to accommodate reals, functions from R to R, and specific functionals like the Riemann intergral. Considering, however, appropriate third-order functions and coding continuous functions by sets of natural numbers, one can carry out this development in Hilbert and Bernays' main formalism which is described next.

[69] The formalism is not only of interest as a framework for mathematical practice in classical analysis, but it has also important metamathematical stability properties. First of all, there are proof-theoretic equivalences to various versions of set theory WITHOUT the power-set axiom; secondly, the formalism with AC is conservative over the formalism with just CA for Π_4^1-sentences. This point has been emphasized by Kreisel. For technical details see Apt and Marek's paper, 'Second order arithmetic and related topics', *Ann. Math. Logic* **6** (1974), 177–229.

[70] It is given as in Feferman's and my first chapter of the *Lecture Notes* volume already mentioned.

[71] Even in the restriction to the purely number-theoretic part of the language and to intuitionistic logic one can define $\neg$, $\vee$, $\leftrightarrow$ from $\rightarrow$, $\wedge$, $\bot$.

[72] Quantifier-free formulas and formulas with bounded number-quantifiers. Π_∞^0-formulas contain only number-quantifiers; Π_1^1-(Σ_1^1-, Σ_2^1-)formulas are of the form $(\forall X)F((\exists X)F, (\exists X)(\forall Y)F)$, where F is arithmetic.

[73] Cp. note 72.

[74] [Feferman, 1964], p. 102.

[75] $(\Pi_\infty^0\text{–CA})$ proves the consistency of Z, and is thus stronger than $(\Pi_\infty^0\text{–CA})\restriction$. For a proof-theoretic argument concerning $(\Sigma_1^1\text{–AC})\restriction$ and Z, compare [Feferman, Sieg] in the LN-volume. This result is originally due to Barwise and Schlipf.

[76] Such calculi had been introduced by Hertz. In [Gentzen, 1934/5] they were investigated and applied to solve some proof-theoretic problems, e.g., consistency of Z^-. Gentzen's method was developed further by, among others, Lorenzen, Schütte, Takeuti, Kreisel, Feferman, and Tait.

[77] Indeed, given the length and cut-rank of $\mathcal{D}$ one can determine a bound on the length of $\mathfrak{E}$. This more detailed information can also be given in the case of the cut-elimination theorem for PL below; there it is crucial.

[78] $\Gamma[G]$ denotes a sequence which consists of the elements of Γ and is interspersed with occurrences of G. Π_1^0-(Σ_1^0-)formulas are of the form $(\forall x)F((\exists x)F)$, where F is quantifier-free.

[79] Thus neither the addition of extra sorts nor that of quantificational logic will allow the proof of new real statements.

[80] Notice that this is "almost" the reflection principle as formulated in (Adeq 2) above; but the partial reflection cannot be improved to $\mathbb{P}r_T(a, {}^\ulcorner F^\urcorner) \rightarrow F$ as long as $\mathfrak{E}$ includes Π_1^0-sentences and T and T^* are equiconsistent.

[81] $\mathrm{ID}_{<\nu}$ is described below. The extension of Gödel and Gentzen's result to these theories is easily proved; using a trick of Friedman's it can be established also for Π_2^0-sentences. The work of Feferman and Friedman yields proof-theoretical equivalences for classes of sentences, which include all arithmetic ones. Detailed statements of theorems and proofs are in [Feferman, Sieg].

[82] Partial truth-definitions and reflection principles are treated carefully in Troelstra's *Metamathematical Investigation of Intuitionistic Arithmetic and Analysis, Lecture Notes in Mathematics* 344, Springer Verlag, Berlin, pp. 33–37.

[83] [Gentzen, 1934/35], p. 416.

[84] Gentzen gave a natural system of notations for the segment of the second number class determined by ϵ_0. One can use it to define (via an appropriate effective coding) a primitive recursive well-ordering of the natural numbers. A precise formulation of

transfinite induction and recursion principles is given in [Tait, 1965]. I denote that theory by PRA + TI(β), if those principles concern a primitive recursive well-ordering of ordinal β.

[85] n is used here as a variable ranging over natural numbers and as a syntactic variable ranging over numerals.

[86] The resulting theory is easily seen to be equivalent to Z.

[87] For simplicity I assume again that F is quantifier-free.

[88] For a beautiful presentation of a quite similar argument see Schwichtenberg's 'Proof theory: some applications of cut-elimination' in *Handbook of Mathematical Logic*, J. Barwise (ed.), North-Holland, Amsterdam, 1977, pp. 867–895.

[89] This is best possible in the sense that (i) PRA + TI(ϵ_0) allows the proof of the consistency of Z, and (ii) PRA + TI(β) is contained in Z for each $\beta < \epsilon_0$. (The latter fact was shown by Gentzen.)

[90] The proof-theoretic work on predicative analysis is described in [Feferman, 1964] and in Tait's *Normal Derivability in Classical Logic*, (*Lecture Notes in Mathematics*, 72), Springer Verlag, Berlin, pp. 204–236. The first steps in analysing impredicative subsystems of analysis were taken by Takeuti.

[91] For an attempt to describe the special features of i.d. classes I refer to [Feferman, Sieg].

[92] $\mu, \nu, \ldots$ are used here as syntactic variables over numerals to indicate that the numerals are related via the well-ordering $<$ to an ordinal of the classical second number class.

[93] Positivity is a purely syntactic concept and taken here in a very broad sense; see [Feferman, Sieg].

[94] The starting point of my work was a paper of Tait's, namely, 'Applications of the cut elimination theorem to some subsystems of classical analysis', in *Intuitionism and Proof Theory*, Kino, Myhill, and Vesley (eds.), Amsterdam, 1970, pp. 475–488.

[95] One can assume without loss of generality that a given derivation in ID$_\nu$ contains references to only one i.d. class.

[96] ρ is again some primitive recursive function. As to the claim that this holds for all negative arithmetic and Π_2^0–sentences, compare note 81.

[97] ID$^i_{<\omega}(\mathcal{O})$ and ID$^i_{\leq\epsilon_0}(\mathcal{O})$ can be interpreted in $(\Pi_1^1\text{–CA})\restriction$, respectively $(\Sigma_2^1\text{–AC})$, such that at least all arithmetic sentences are preserved. Consequently, these theories are equiconsistent.

[98] Troelstra, 'Extended bar induction of type zero,' in *The Kleene Symposium*, Amsterdam, 1980, pp. 237–316; Feferman and Sieg, 'Proof theoretic equivalences between classical and constructive theories for analysis', *LN*-volume by Buchholz, et al.

[99] An example of a striking result (but not established with reductions described here) is that obtained by Kreisel and Troelstra, showing the proof-theoretic equivalence of ID$_1$, their system CS for intuitionistic analysis with principles for choice sequences, and Kleene's theory for intuitionistic analysis. (in *Annals of Mathematical Logic*, 1 (1970), 229–387, and Ibid. 3 (1971), 437–439.)

[100] This suggestion clearly leaves untouched the treatment of the continuum in Dedekind's classical way using the notion of the powerset of N. For a detailed "defense" of this geometrically motivated approach see [Bernays, 1978], pp. 6–9.

[101] Indeed, uses of the notion within mathematics should be looked for.

[102] K. Gödel, 'On an extension of finitary mathematics which has not yet been used'; mimeographed translation of Gödel's paper in *Dialectica*, **12**, (1958), 280–287, with additional notes. The quotation is from p. 16.

[103] Brouwer's phrase is from *Collected Works* I, A. Heyting (editor), Amsterdam, 1975, p. 412; recall, that for the Hilbert school finitist and intuitionistic mathematics coincided.

[104] The use of classical logic can be seen to be inessential (at least for proofs of negative arithmetic and Π_2^0-sentences) by metamathematical means. But it should be mentioned here that E. Bishop in his *Foundations for Constructive Analysis* developed analysis in an almost Kroneckerian spirit. Work of Friedman and Feferman shows that Bishop's constructive mathematics can be carried out in conservative extensions of intuitionistic number theory. For references to this area of foundational investigations see S. Feferman, 'Constructive theories of functions and classes', in *Logic Colloquium '78*, Boffa, van Dalen, McAloon (eds.), Amsterdam, 1979, pp. 159–224.

[105] Brouwer's *Collected Works*, I, p. 85.

[106] See [Gödel, 1944], p. 219 in the reprinted version published in *Philosophy of Mathematics*, P. Benacerraf and H. Putnam (eds.), 1964.

[107] And thus the condition that something has been proved is not intuitively determined (anschaulich bestimmt). These considerations are found in "Über den Platonismus in der Mathematik", in [Bernays, 1976], p. 71.

[108] [Brouwer, 1927], p. 460, fn. 8.

[109] Ibid., p. 460, fn. 8.

[110] [Zermelo, 1931], p. 85; cp. also the newly published documents from Zermelo's *Nachlass* in [Moore, 1980].

[111] [Zermelo, 1935], pp. 144–145. To the contemporary work on infinitary logics I can only allude; an introductory survey is given by Jon Barwise, 'Infinitary Logics', in *Modern Logic – A Survey*, Agazzi (ed.), D. Reidel, Dordrecht, 1981, pp. 93–112.

[112] The truth of initial statements and the correctness of rules in certainly assumed.

[113] [Gödel, 1931], pp. 28–29, fn. 48a.

[114] Gödel discussed in his 1958 paper, referred to in note 102, a particular extension of finitist mathematics and some general points concerning such extensions. Gentzen suggested to use progressively longer segments of the second number class for consistency proofs, in the form of appropriate systems of ordinal notions.

[115] [Reid, 1970], p. 218; cp. also Gödel's remarks quoted in Hao Wang's *From Mathematics to Philosophy*, London, 1974, pp. 325–326.

[116] By that one may simply mean that the operation is constructive and can be grasped "directly" for example, successor operation and the joining of an effectively given sequence of trees in the definition of $\mathcal{O}$.

[117] Bernays suggests to view mathematics as "die Wissenschaft von den idealisierten Strukturen"; such idealized structures are in a complex way integrated in our broad intellectual experience. "Wenn wir die zuvor dargelegte Auffassung zugrunde legen, wonach die Mathematik die Wissenschaft von den idealisierten Strukturen ist, so haben wir damit für die Grundlagenforschung der Mathematik eine Haltung, welche uns vor übersteigerten Aporien und vor forcierten Konstruktionen bewahrt und welche auch nicht angefochten wird, wenn die Grundlagenforschung vieles Erstaunliche zutage bringt." [Bernays, 1976], p. 188. This notion invites philosophical reflection on two general questions: (1) what is the status of idealized structures? (2) what are the grounds

on which infinitistic principles concerning them are assumed?
[118] [Bernays, 1976], p. 78.

REFERENCES

Bernays, P.: 1922A, 'Über Hilberts Gedanken zur Grundlegung der Arithmetik', *Jahresbericht* DMV **31**, 10–19.

Bernays, P.: 1922B, 'Die Bedeutung Hilberts für die Philosophie der Mathematik', *Die Naturwissenschaften*, Heft **4**, 93–99.

Bernays, P.: 1930, 'Die Philosophie der Mathematik und die Hilbertsche Beweistheorie', in *Abhandlungen zur Philosophie der Mathematik*; see Bernays, 1976, pp. 17–61.

Bernays, P.: 1935, 'Hilberts Untersuchungen über die Grundlagen der Arithmetik', in *Gesammelte Abhandlungen*; see Hilbert, 1935, pp. 196–216.

Bernays, P.: 1950, 'Mathematische Existenz und Widerspruchsfreiheit', in *Abhandlungen zur Philosophie der Mathematik*; See Bernays, 1976, pp. 92–106.

Bernays, P.: 1967, 'David Hilbert', in P. Edwards (ed.), *The Encyclopedia of Philosophy* III, New York, pp. 496–504.

Bernays, P.: 1976, *Abhandlungen zur Philosophie der Mathematik*, Darmstadt.

Bernays, P.: 1978, 'Bemerkungen zu Lorenzens Stellungnahme in der Philosophie der Mathematik', in K. Lorenz (ed.), *Konstruktionen versus Positionen*, Berlin, pp. 3–16.

Borel, E.: 1914, *Lecons sur la théorie des fonctions*, Deuxième édition, Paris.

Brouwer, L. E. J.: 1927, 'Über Definitionsbereiche von Funktionen', in J. van Heijenoort (ed.), *From Frege to Gödel: A Source Book in Mathematical Logic*, 1879–1931, Harvard University Press, Cambridge, pp. 446–463.

Buchholz, W., S. Feferman, W. Pohlers, W. Sieg: 1981, 'Iterated Inductive Definitions and Subsystems of Analysis: Recent Proof-Theoretical Studies', *Springer Lecture Notes in Mathematics* **897**, Berlin, Heidelberg, New York.

Dedekind, R.: 1932, *Gesammelte Mathematische Werke* (Bd. III), Braunschweig.

Feferman, S.: 1964, 'Systems of Predicative Analysis', *Journal of Symbolic Logic* **29**, 1–30. Reprinted in J. Hintikka (ed.): 1969, *The Philosophy of Mathematics*, Oxford, pp. 95–127.

Feferman, S.: 1970, 'Formal Theories for Transfinite Iterations of Generalized Inductive Definitions and Some Subsystems of Analysis', in Kino, Myhill, and Vesley (eds.), *Intuitionism and Proof Theory*, Amsterdam, London, pp. 303–325.

Feferman, S.: 1977, 'Theories of Finite Type Related to Mathematical Practice', in J. Barwise (ed.), *Handbook of Mathematical Logic*, Amsterdam, pp. 913–971.

Feferman, S: 1978, 'A More Perspicuous Formal System for Predicativity', in K. Lorenz (ed.), *Konstruktionen Versus Positionen*, Berlin, pp. 68–93.

Feferman, S. and W. Sieg: 1981, 'Iterated Inductive Definitions and Subsystems of Analysis', in W. Buchholz, S. Feferman, W. Pohlers, and W. Sieg, 'Iterated Inductive Definitions and Subsystems of Analysis: Recent Proof-Theoretical Studies', *Springer Lecture Notes in Mathematics* **897**, Berlin, Heidelberg, New York, pp. 16–77.

Friedman, H.: 1970, 'Iterated Inductive Definitions and Σ_2^1-AC', Kino, Myhill, and Vesley (eds.), *Intuitionism and Proof Theory*, Amsterdam, London, pp. 435–442.

Friedman, H.: 1976, 'The Arithmetic Theory of Sets and Functions I', (mimeographed).

Gandy, R. O.: 1973, 'Bertrand Russell, as Mathematician', *Bulletin of the London*

Mathematical Society 5 pp. 342–348.

Gauss, K. F. and F. W. Bessel: 1880, *Briefwechsel*, Leipzig.

Gentzen, G.: 1934/5, 'Untersuchungen über das logische Schließen I, II', *Math. Zeitschrift* 39, 176–210, 405–431.

Gödel, K.: 1931, 'Über formal unentscheidbare Sätze der Principia Mathematica und Verwandter Systeme I', in M. David (ed.), *The Undecidable*, Hewlett, New York, pp. 5–38.

Gödel, K.: 1944, 'Russell's Mathematical Logic', in P. A. Schilpp (ed.), *The Philosophy of Bertrand Russell*, Evanston, pp. 125–153.

Hilbert, D.: 1900, 'Über den Zahlbegriff', *Jahresbericht DMV* 8 pp. 180–194.

Hilbert, D.: 1902, 'Sur les problèmes futurs des mathématiques', in *Compte Rendu du Deuxième Congrès International des Mathématiciens*, Paris, pp. 58–114.

Hilbert, D.: 1904, 'Über die Grundlagen der Logik und Arithmetik', in *Grundlagen der Geometrie*, 5. Auflage, Leipzig and Berlin, pp. 243–258.

Hilbert, D.: 1922, *Grundlagen der Geometrie*, 5. Auflage, Leipzig and Berlin.

Hilbert, D.: 1928, 'Probleme der Grundlegung der Mathematik', in K. Reidemeister (ed), *Hilbert*, Berlin, pp. 9–19.

Hilbert, D.: 1931, 'Die Grundlegung der Elementaren Zahlenlehre', *Math. Annalen* 104, 485–494.

Hilbert, D.: 1935, *Gesammelte Abhandlungen*, 3. Band, Berlin.

Hilbert, D. and P. Bernays: 1934 (1968): *Grundlagen der Mathematik* 1, Zweite Auflage, Berlin.

Kreisel, G.: 1968, 'Survey of Proof Theory', *Journal of Symbolic Logic* 33, 321–388.

Kronecker, L.: 1887, 'Über den Zahlbegriff', in *Werke* III, pp. 251–274.

Kronecker, L.: 1901, *Vorlesungen zur Zahlentheorie*, (K. Hensel, editor), Leipzig.

Moore, G. H.: 1980, 'Beyond First-Order Logic: The Historical Interplay Between Mathematical Logic and Axiomatic Set Theory', *History and Philosophy of Logic* 1, 95–137.

Poincaré, H.: 1902, 'Du rôle de l'Intuition et de la Logique en Mathématiques' in *Compte Rendu du Deuxième Congrès International des Mathématiciens*, Paris, 115–130.

Reid, C.: 1970, *Hilbert*, Berlin.

Russell, B.: 1908, 'Mathematical Logic as Based on the Theory of Types', in J. van Heijenoort (ed.), *From Frege to Gödel: A Source Book in Mathematical Logic 1879–1931*, Harvard University Press, Cambridge, pp. 150–182.

Sieg, W.: 1977, *Trees in Metamathematics (Theories of Inductive Definitions and Subsystems of Analysis)*, Ph.D. Thesis, Stanford.

Tait, W. W.: 1965, 'Functionals Defined by Transfinite Recursion', *Journal of Symbolic Logic* 30, 155–174.

Takeuti, G.: 1978, *Two Applications of Logic to Mathematics*, Princeton University Press, Princeton.

van Heijenoort, J.: 1967, *From Frege to Gödel: A Source Book in Mathematical Logic 1879–1931*, Harvard University Press, Cambridge.

Weber, H.: 1893, 'Leopold Kronecker', *Jahresbericht DMV*2, 5–31.

Weyl, H.: 1918, *Das Kontinuum*, Leipzig.

Whitehead, A. N. and B. Russell: 1910, *Principia Mathematica* I, Cambridge.

Zermelo, E.: 1908, 'Untersuchungen über die Grundlagen der Mengenlehre I', *Math. Ann.* 65, 261–281.

Zermelo, E.: 1931, 'Über Stufen der Quantifikation und die Logik des Unendlichen', *Jahresbericht DMV* **31**, 85–88.

Zermelo, E.: 1935, 'Grundlagen einer Allgemeinen Theorie der Mathematischen Satzsysteme', *Fundamenta Mathematicae* **25**, 136–146.

Dept. of Philosophy
Columbia University
New York, NY 10027
U.S.A.

RAYMOND M. SMULLYAN

CHAMELEONIC LANGUAGES

INTRODUCTION

We would like to open with a report of a delightful remark from a correspondent who wrote: "I have heard of your chameleonic languages. I do not know what they are, except that I assume they are not what they appear to be."

Now, there are certain words in the English language whose denotation is dependent on the surrounding context, and which might accordingly be labeled *chameleonic*. (*Indexical* is the more usual term, but the more colorful word *chameleonic* seems a bit more appropriate for this article.) For example, the word *now* denotes different instants of time when uttered at different times; the word *I* designates different people when used by different persons[1]; the word *you* designates the person to whom it is addressed. Like a chameleon whose color depends on its surroundings, these words change their denotation from context to context. A classical example of a chameleonic term, and one which appears in the well-known formulation of the liar paradox, is the word *this* – in such contexts as "This sentence has property *P*." Such a sentence might be termed *self-referential*, in that it expresses the proposition that it itself has the property *P*, and is accordingly true if and only if it does have the property *P*. Now, the usual first-order arithmetic theories, in which one is interested in obtaining incompleteness and undecidability results, do not contain chameleonic symbols. (One gets around this by the use of Gödel's diagonal function [2], or for some languages, the simpler norm function introduced in [3]). In this article, instead of circumventing the chameleonic term, we formalize it directly.

In Part 1 we work in informal English. Given a property *P* of sentences, we show how to construct a nonindexical sentence (a *normal* sentence, as we call it) which says that it has property *P*. Then we consider the problem of *cross-reference* – given two properties P_1, P_2, how to construct normal sentences S_1, S_2, such that S_1 says that S_2 has property P_2, and S_2 says that S_1 has property P_1. (This is related to the

Synthese **60** (1984) 201–224.   0039–7857/84/0602–0201 $02.40
© 1984 *by D. Reidel Publishing Company*

doubly Gödelian islands of [4].) Then we consider the problem of *simultaneous self-reference* – given two binary relations $R_1(X, Y)$, $R_2(X, Y)$ on sentences, how to construct normal sentences S_1, S_2, such that S_1 says that $R_1(S_1, S_2)$, and S_2 says that $R_2(S_1, S_2)$. (This is related to the author's double-recursion theorem [5], or to the closely related generalized diagonal lemma of Boolos [1].) We have found two quite different methods of achieving cross-reference and simultaneous self-reference: the first uses two indexical terms "this" and "that"; the second uses only the one indexical term "this". The first method when formalized in arithmetic yields essentially the construction of Boolos [1]; the second yields results which are more novel (Theorems 5, 6, 7, 8, 9 in Part 2).

In Part 2 we arithmetize some of the constructions of Part 1. The gist of the idea is that we take a first-order arithmetic theory (T), add a new individual symbol "σ" (the *chameleonic* symbol) and in any sentence C in which "σ" occurs (a "chameleonic" sentence, as we call it), the symbol "σ" is interpreted as denoting the Gödel number of the entire sentence C. (Thus these chameleonic sentences refer to their own Gödel numbers in as direct a manner as can be imagined.) Each chameleonic sentence C is assigned an ordinary sentence called the *translation* of C, which is true if and only if C is chameleonically true (true, that is, when σ is interpreted as being the Gödel number of C). Given a subset P of the set of true nonchameleonic sentences (whose elements are the *provable* sentences, say in Peano Arithmetic), we let P^σ be the set of all chameleonic sentences whose translation is in P. Both the set T^σ (the set of chameleonically true sentences) and the set P^σ have bizarre properties; they are neither consistent nor closed under modus ponens. (Though they are inconsistent, no single member is inconsistent.) Despite their bizarre properties, these sets are far from useless – for example, one way to get an undecidable sentence of the original theory is to represent in it the set of Gödel numbers of the set P^σ. This method becomes virtually Gödel's diagonal method if we replace "σ" by a variable. It is no simpler than the standard method, but is heuristically illuminating.

After carrying out various "multiply self-referential" constructions in arithmetic theories, we concluded with a consideration of the following problem: A chameleonic formula $K(x,\sigma)$ (which has x as the only free variable, but also may have the chameleonic symbol σ) can be said to represent a set – namely the set of all n such that the chameleonic sentence $K(\bar{n}, \sigma)$ is chameleonically true. Such sets are perfectly well defined, but are they all necessarily representable in the original theory? We know of no reason why in general they should be, but we give a sufficient condition that they all are, and this condition does apply to all theories in which all recursive functions are definable.

PART 1 – CHAMELEONIC CONSTRUCTIONS FOR NATURAL LANGUAGES

1. We begin with a heuristic sketch of several of the main ideas involved. Let us consider ordinary English, augmented with some means of naming expressions – say the familiar device of quotation. Suppose we wish to construct a sentence which asserts that it itself has a certain property J – say the property that John is reading it. We can do this by writing down the following:

(1)   John is reading this sentence.

It is understood that in (1) the indexical phrase "this sentence" denotes the whole of Sentence (1), and so (1) is true if and only if John is reading (1).

We now wish to construct a sentence which does not use any indexical phrase, but which asserts that John is reading it. Let us define the *translation* of a sentence X to be the result of substituting the name of X for every occurrence of "this sentence" in X. For example, the translation of (1) is the following sentence (1)*.

(1)*   John is reading "John is reading this sentence."

We would not call (1)* a chameleonic sentence, as we would (1), because although the indexical phrase "this sentence" occurs in it, it occurs only in a context which is enclosed in quotation marks – thus it is not *used* but only mentioned. (Indeed, we could use some other device for naming the expression such as spelling – for example, we could use the symbols "L_1", ..., "L_{26}" as respective names of the letters "A", ..., "Z", and "L_{27}" as a name of a space between letters. Then the translation of (1) would consist of "John is reading the translation of" followed by the name of (1), and this name would consist of a string of the symbols "L_1", ..., "L_{27}", and the word "this" wouldn't occur in it at all.) By contrast, the word "this" is *used* but not mentioned in (1).

And so we would call (1) a *chameleonic* sentence, and (1)* a *normal* sentence.

However, (1)* does not say that John is reading (1)*; it says that John is reading (1). And so we have not yet succeeded in constructing a *normal* self-referential sentence.

Let us make another try: consider the following chameleonic sentence:

(2)   John is reading the translation of this sentence.

Its translation is

(2)*   John is reading the translation of "John is reading the translation of this sentence."

Now, (2)* asserts that John is reading the translation of (2), but the translation of (2) is (2)*. Therefore (2)* asserts that John is reading the very sentence (2)*.

We might note that the sentences (2), (2)* are equivalent; they both say that John is reading (2)*, yet (2)* is self-referential but (2) is not. We might also note the following fact: Suppose we were not told what was meant by the translation of a sentence; all we were told was that there was a certain function which assigned to every chameleonic sentence S a normal sentence S^* called the *translation* of S such that S^* is true if and only if S is true. Then we would not know what the translation of (2) actually was, but we would still know that this translation, whatever it was, must be true if and only if John is reading it. This suggests that an abstract theory of chameleonic translations might be of interest, and we plan to pursue this elsewhere.

2. Next we wish to point out some bizarre properties of *truth* when applied to chameleonic sentences. Let us consider the following two chameleonic sentences:

(1)   This sentence has property P.
(2)   This sentence has property Q.

Until we have assigned actual properties P, Q to the symbols "P" and "Q", we cannot assign truth values to (1) or (2). The first curious thing to observe is that it is possible to assing the *same* property to the symbols "P", "Q" in such a manner that the truth-values of (1), (2) are different: For any expression X, define $P(X)$ to mean that X contains at least one occurrence of "P" and $Q(X)$ to also mean that X contains at least one occurrence of "P". Under this interpretation, (1) is true but (2) is false (since (1) does contain an occurrence of "P" and (2) does not).

Let us now use the symbol "σ'" as an abbreviation of "this sentence", and also abbreviate $\ulcorner x$ has property $P\urcorner$ by $\ulcorner Px\urcorner$ and $\ulcorner x$ has property $Q\urcorner$ by $\ulcorner Qx\urcorner$. So we rewrite (1), (2) thus:

(1) $P\sigma$
(2) $Q\sigma$

Now we consider all sentential combinations of (1) and (2) – i.e., all sentences built from (1), (2) by using the logical connectives. In any such sentence X, the symbol "σ" will be understood as designating X itself. For example in (1), 'σ' designates "$P\sigma$" and in (2), "σ" designates "$Q\sigma$". Let us now look at the following sentence:

(3) $P\sigma \wedge Q\sigma$

Now, if we assign properties P, Q to the symbols "P", "Q", it is *not* necessarily the case that (3) is true if and only if (1) and (2) are both true, because (1) is true if and only if (1) has property P; (2) is true if and only if (2) has property Q, but (3) is true if and only if (3) has both property P and property Q. Indeed, we can assign values to "P", "Q" in such a manner that $P\sigma \wedge Q\sigma$ is true, but "$P\sigma$" and "$Q\sigma$" are both false: Define $P(X)$ to mean that X contains both symbols "P", "Q", and define $Q(X)$ to mean the same thing. Under this interpretation, "$P\sigma$" is false (since it doesn't contain "Q") and "$Q\sigma$" is false (since it doesn't contain "P"). yet $P\sigma \wedge Q\sigma$ is true. Again, we could give values to "P", "Q" such that (1), (2) are both true, yet (3) is false: Define $P(X)$ and $Q(X)$ to mean that X lacks at least one of the symbols "P", "Q" – or alternatively, define $P(X)$ and $Q(X)$ to mean that X contains exactly two symbols. Under either of these interpretations, (1), (2) are both true, but (3) is false.

Negation also behaves in a nonstandard fashion. Consider the following sentence

(4) $\sim\sim P\sigma$

Under a given interpretation of "P", the sentence (1), (4) are not necessarily equivalent. For example, take $P(X)$ to mean that X contains at least one occurrence of "$\sim$". Under this interpretation, (1) is false, but (4) is true.

One can also interpret "P" in such a manner that the sentences "$P\sigma$" and "$\sim P\sigma$" are both true (define $P(X)$ to mean that X contains no occurrence of "$\sim$"). One can also interpret "P" such that neither of the sentences "$P\sigma$", "$\sim P\sigma$" are true: Define $P(X)$ to mean that X *does* contain at least one occurrence of "$\sim$".

3. *Cross Reference* (*First Method*). We now consider two properties – say the property of being read by John and the property of being read by Paul. We wish to construct two normal sentences S_1, S_2 such that S_1 says that John is reading S_2 and S_2 says that Paul is reading S_1. We have two different methods of doing this; the first uses two chameleonic words – "this" and "that". Now, if a person utters a phrase involving the indexical words "you" and "I", the sentence does not have any truth value in itself; it only has a truth value relative to the person addressed. Similarly, given a chameleonic sentence C involving the phrases "this sentence" and "that sentence", C has no truth value in itself, but we define it to be true *relative to a given expression X* if C is true when "this sentence" is interpreted as denoting C and "that sentence" is interpreted as denoting X. For example, consider the following two chameleonic sentences:

C_1:   John is reading that sentence.
C_2:   Paul is reading that sentence.

Neither C_1 nor C_2 have independent truth values, but C_1 is true relative to C_2 if and only if John is reading C_2, and C_2 is true relative to C_1 if and only if Paul is reading C_1. And so we have a "cross-referential" pair (C_1, C_2). But C_1, C_2 are both chameleonic. What we want is a cross-referential pair (S_1, S_2) of *normal* sentences.

Well, given a chameleonic sentence C containing one or both of the chameleonic phrases "this sentence", "that sentence", and given an expression X, we define the translation of C *with respect to X* to be the sentence obtained from C by substituting the name of C for every occurrence of the phrase "this sentence" and the name of X for every occurrence of the phrase "that sentence". The translation of C with respect to X is a normal sentence, and is a true sentence if and only if C is true relative to X.

Now consider the following two chameleonic sentences:

(1)   John is reading the translation of that sentence with respect

to this sentence.

(2)   Paul is reading the translation of that sentence with respect to this sentence.

Let S_1 be the translation of (1) with respect to (2) and S_2 be the translation of (2) with respect to (1). S_1 is true if and only if (1) is true relative to (2), which in turn is the case if and only if John is reading the translation of (2) with respect to (1). But the translation of (2) with respect to (1) is S_2. Therefore, S_1 is true if and only if John is reading S_2. similarly, S_2 is true if and only if Paul is reading S_1.

We can, of course, display the sentences S_1, S_2 explicitly:

S_1:   John is reading the translation of "Paul is reading the translation of that sentence with respect to this sentence." with respect to "John is reading the translation of that sentence with respect to this sentence.".

S_2:   Like S_1, interchanging "John" with "Paul".

Exercise – Using three chameleonic phrases "this sentence", "that sentence", "the other sentence", define an appropriate notion of the translation of *S with respect to X and Y* and construct normal sentences S_1, S_2, S_3 such that S_1 says that John is reading S_2; S_2 says that Paul is reading S_3, and S_3 says that Gertrude is reading S_1.

4. *Cross Reference Using Only One Chameleonic Term*. The construction which now follows will have its arithmetic analogue in Part 2.

Consider the following chameleonic sentence:

(1)   John is reading the sentence consisting of "Paul is reading" followed by the name of the translation of this sentence.

Let S_1 be the translation of (1). It says that John is reading the sentence consisting of "Paul is reading" followed by the name of S_1; let S_2 be that sentence. Then S_1 says that John is reading S_2. Since S_2 is the expression consisting of "Paul is reading" followed by the name of S_1, then S_2 says that Paul is reading S_1.

The sentences S_1, S_2 can of course be written down explicitly.

Exercise – Using only one indexical phrase "this sentence", construct three normal sentences S_1, S_2, S_3, such that S_1 says that John is reading S_2; S_2 says that Paul is reading S_3, and S_3 says that Gertrude is reading S_1.

5. *Simultaneous Self-Reference*. Given two binary relations $R_1(X, Y)$ and $R_2(X, Y)$ on sentences, we wish to construct normal sentences S_1, S_2 such that S_1 asserts that $R_1(S_1, S_2)$ and S_2 asserts that $R_2(S_1, S_2)$. Suppose, for example, we want to construct normal sentences S_1, S_2 such that S_1 says that John prefers S_1 to S_2, and S_2 says that Paul prefers S_1 to S_2.

Again we can treat this in two quite different ways; the first (whose arithmetic counterpart) is similar to Boolos's construction [1]; the second (which is along more novel lines) will be treated in the next section.

For the first method, we first note that it is obvious how to construct *chameleonic* sentences C_1, C_2 such that C_1 is true *relative to C_2* if and only if John prefers C_1 to C_2, and C_2 is true *relative to C_1* if and only if Paul prefers C_1 to C_2 – namely, take the following:

C_1:   John prefers this sentence to that sentence.
C_2:   Paul prefers that sentence to this sentence.

We are interested, however, in constructing *normal* sentences S_1, S_2 which work. Well, we first consider the following chameleonic sentences:

(1)   John prefers the translation of this sentence with respect to that sentence to the translation of that sentence with respect to this sentence.

(2)   Paul prefers the translation of that sentence with respect to this sentence to the translation of this sentence with respect to that sentence.

We let S_1 be the translation of (1) with respect to (2) and S_2 to be the translation of (2) with respect to (1). To write down S_1, S_2 explicitly would be a bit of a mess, but fortunately we don't have to do this in order to see that the pair (S_1, S_2) works: S_1 is true if and only if (1) is true relative to (2), which in turn is the case if and only if John prefers the translation of (1) to (2) – i.e., if and only if John prefers S_1 to S_2. Similarly, S_2 is true if and only if Paul prefers S_1 to S_2.

6. *Simultaneous Self Reference Using Only One Chameleonic Term*. Now, how can we do the same thing using only the one chameleonic term "this"? This is not so obvious. Indeed, it is not even

immediately obvious how to construct two *chameleonic* sentences C_1, C_2 (using only the chameleonic term "this") such that C_1 is chameleonically true if and only if John prefers C_1 to C_2, and C_2 is chameleonically true if and only if Paul prefers C_1 to C_2. (When we say that a chameleonic sentence C is "chameleonically true", we of course mean true when "this sentence" is interpreted as referring to C.)

First of all, if we are using quotation as our means of naming expressions, then we must modify our previous definition of "translation" as follows: Call an occurrence of an expression Y in an expression X a *quotation free* occurrence if it is not part of any part of X which is enclosed in quotes. Then we must define the *translation* of X to be the result of substituting the quotation of X *for every quotation free occurrence* of "this sentence" in X. (For the previous constructions, this distinction didn't matter; for those that follow, it does. Also, if we named expressions by spelling instead of quotation, this point needn't arise.)

Now, consider the following chameleonic sentence:

C_1:   John prefers this sentence to the sentence consisting of "Paul prefers" followed by the name of this sentence, followed again by "to this sentence.".

The chameleonic sentence C_1 says that John prefers C_1 to the chameleonic sentence C_2, where C_2 can be constructed as follows: First write down "Paul prefers", then follow it by the name (quotation) of the sentence C_1, and then follow that by "to this sentence." Fully written out, we get the following:

C_2:   Paul prefers "John prefers this sentence to the sentence consisting of "Paul prefers" followed by the name of this sentence, followed again by "to this sentence."" to this sentence.

To see the construction more clearly, let $\bar{C}_1$ be the quotation of the sentence C_1, and let us use the symbol "$\bar{C}_1$" to *abbreviate* the quotation of C_1. Then the following expression *abbreviates* C_2:

Paul prefers $\bar{C}_1$ to this sentence.

It is now clear that C_2 says that Paul prefers C_1 to C_2 (because in C_2, "this sentence" is understood as denoting C_2). And, as we have seen, C_1 says that John prefers C_1 to C_2.

This solves the problem of constructing a *chameleonic* pair (C_1, C_2) that works. But we wish to construct a *normal* pair (S_1, S_2) that works. We do this by starting out with the following chameleonic sentence D_1 instead of C_1:

> D_1:   John prefers the translation of this sentence to the sentence consisting of "Paul prefers" followed by the name of the translation of this sentence, followed again by "to the translation of this sentence.".

We let S_1 be the translation of D_1. (To write it out in full is a rather horrendous task!) The sentence S_1 says the John prefers the sentence S_1 to the translation of the chameleonic sentence D_2, where D_2 is obtained by writing down "Paul prefers", followed by the quotation of S_1, followed again by "to the translation of this sentence." We shall not display D_2 explicitly; we use "$\bar{S}_1$" to abbreviate the quotation of the (horrendous) sentence S_1, and the following expression *abbreviates D_2*:

> John prefers $\bar{S}_1$ to the translation of this sentence.

We let S_2 be the translation of D_2. Letting "$\bar{D}_2$" abbreviate the quotation of the chameleonic sentence D_2, the following expression *abbreviates S_2*.

> Paul prefers $\bar{S}_1$ to the translation of $\bar{D}_2$.

Thus S_2 says that Paul prefers S_1 to the translation of D_2 – in other words that Paul prefers S_1 to S_2. And S_1 says that John prefers S_1 to the translation of D_2 – in other words that John prefers S_1 to S_2.

Of course the former construction (using two chameleonic terms) is a good deal simpler (and more natural) than the construction above. However, the problem of whether simultaneous self-reference could be obtained using only one chameleonic term intrigued us in its own right, and though we had no applications in mind, it led us to some of the results in arithmetic theories in Part 2 which we could not have obtained from the former construction.

Before turning to Part 2 we would like to remark that although all the constructions of Part 1 have been carried out for the admittedly imprecise language of English, it is possible to completely formalize them – either for a well-defined portion of English (for which an exact truth definition can be given), or for precise symbolic language formalizing either quotation or spelling, and taking the translation function, the naming function and the concatenation function as primitive.

PART 2 – CHAMELEONIC ARITHMETIC LANGUAGES

7. *Preliminaries.* We consider a first-order arithmetic theory (T) with standard formalization in the sense of Tarski[7]. Each natural number n has a name in the theory; we let $\bar{n}$ be this numeral which names n. (Tarski uses "Δ_n" for "$\bar{n}$".) We now construct the chameleonic extension $(T)^\sigma$ of (T) as follows: We take a new symbol "σ" and adjoin it as an individual symbol; it is neither a variable nor a numeral; we call it the *chameleonic* symbol. By a *σ-formula* we mean an expression like a formula, except that "σ" can occur in place of variables or numerals. Alternatively a σ-formula is any expression resulting from a formula by substituting "σ" for all free occurrences of some variable. What are ordinarily called "formulas" we will sometimes call *normal formulas* for emphasis. We shall regard normal formulas as special cases of σ-formulas (namely σ-formulas in which "σ" does not occur). By a σ-sentence we mean a σ-formula containing no free variables. By a *sentence* of (T^σ) we mean either a σ-sentence or a normal sentence (sentence of (T)). For any normal formula $\phi(x)$, by $\phi(\sigma)$ we mean the result of substituting "σ" for every free occurrence of the variable x. If x is the only free variable of $\phi(x)$, then $\phi(\sigma)$ is a σ-sentence. Any σ-sentence is of the form $\phi(\sigma)$, for some formula $\phi(x)$ in which x is the only free variable.

We now take a 1–1 Gödel numbering g of all σ-sentences *onto* the set of positive integers. (*Onto*, of course, means that every positive integer is the Gödel number of some σ-sentence. We do this as a matter of technical convenience; all the proofs that follow can be modified for Gödel numberings which are not onto.) For any σ-sentence $\phi(\sigma)$ and for any n, by $\phi(\bar{n})$ we mean the result of substituting the numeral $\bar{n}$ for every occurrence of "σ" in $\phi(\sigma)$ (or equivalently, substituting $\bar{n}$ for every *free* occurrence of the variable x in $\phi(x)$. By the *translation* of $\phi(\sigma)$ we mean $\phi(\bar{n})$, where n is the Gödel number of $\phi(\sigma)$.

We let T be the set of valid sentences of the theory (T). By T^σ we mean the set of all σ-sentences whose translation is in T. If T is a complete consistent theory, then we might call the elements of T the *true* sentences of the theory (T) and we might call T^σ the *chameleonic* truth set. This set T^σ has all the bizarre properties mentioned in Part 1; the conjunction of two chameleonic sentences C_1, C_2 might be in T^σ even though one or both of them are not in T^σ. Also we might have a chameleonic sentence C such that C, $\sim C$ are both in T^σ. (However $(C \wedge \sim C)$ cannot be in T^σ, since its translation cannot be in T; indeed no logically contradictory sentence can be in T^σ; the set T^σ as a whole is a logically contradictory set, but no single element of it can be a logical contradiction.) Also, we can have a chameleonic sentence of the form $(\exists y)\psi(y, \sigma)$ which is in T^σ, yet none of the sentences $\psi(0, \sigma)$, $\psi(1, \sigma), \ldots, \psi(n. \sigma), \ldots$ are in T^σ. As an example, suppose the Gödel numbering g has the property (which standard Gödel numberings have) that the Gödel number of a sentence is larger than any n for which the numeral $\bar{n}$ appears in the senence. Then none of the chameleonic sentences $\bar{0} = \sigma$, $\bar{1} = \sigma, \ldots, \bar{n} = \sigma, \ldots$ can be true, yet the chameleonic sentence $(\exists y)(y = \sigma)$ is true (its translation is $(\exists y)(y = \bar{n})$, where n is the Gödel number of $(\exists y)(y = \sigma)$, and this translation is obviously a true normal sentence).

8. *Self-reference.* A set A (for natural numbers) is called *representable* in (T) if there is a normal formula $\phi(x)$ such that for all n, $\phi(\bar{n}) \in T$ if and only if $n \in A$ (and such a formula $\phi(x)$ is said to represent A). We call a normal sentence X a *Gödel sentence* for A if either $X \in T$ and $g(X) \in A$, or $X \notin T$ and $g(x) \notin A$ – in other words, $X \in T \leftrightarrow g(x) \in A$. (By $g(X)$ we of course mean the Gödel number of X.) We shall also call a chameleonic sentence C a *chameleonic Gödel* sentence for A if $C \in T^\sigma \leftrightarrow g(C) \in A$.

The following theorem holds for any theory (T) whatsoever; not just those in which certain key functions are definable.

THEOREM 1. For any set A representable in (T), there is a chameleonic Gödel sentence C.

Proof. Let $\phi(x)$ be a normal formula which represents A. Then $\phi(\sigma)$ is a chameleonic Gödel sentence for A, because $\phi(\sigma) \in T^\sigma$ if and only if its translation $\phi(\bar{h}) \in T$, which in turn is the case if and only if $h \in A$. But h is the Gödel number of $\phi(\sigma)$.

For any set W of expressions, we let $g(W)$ be the set of Gödel numbers of the elements of W. For any set A of natural numbers, we let $\bar{A}$ be its complement.

COROLLARY 1. The set $\overline{g(T^\sigma)}$ is not representable in (T).

Proof. If it were, then by Theorem 1 there would be a chameleonic Gödel sentence C for $\overline{g(T^\sigma)}$, and we would have $C \in T^\sigma \leftrightarrow g(C) \in \overline{g(T^\sigma)}$, but $g(C) \in \overline{g(T^\sigma)} \leftrightarrow g(C) \notin \overline{g(T^\sigma)} \leftrightarrow C \notin T^\sigma$, and we would have $C \in T^\sigma \leftrightarrow C \notin T^\sigma$, which is a contradiction.

COROLLARY 2. If (T) is a complete consistent theory, then $g(T^\sigma)$ is not represented in (T).

Proof. If (T) is complete and consistent, then the complement of every set representable in (T) is also representable in (T). (If $\phi(x)$ represents A in (T), then $\sim\phi(x)$ represents $\bar{A}$ in (T).) Result then follows by Corollary 1.

We now see that for *any* complete consistent theory (T), the set of Gödel numbers of the provable sentences of the chameleonic extension of (T) is not representable in the original (T).

8.1 For any number set A and any function $f(x)$ from the set of natural numbers to the set of natural numbers, by $f^{-1}(A)$ is meant the set of all n such that $f(n) \in A$. We shall call f *admissible* (for sets) if for every representable set A, the set $f^{-1}(A)$ is also representable.

For any n, we define $t(n)$ to be the Gödel number of the translation of that chameleonic sentence whose Gödel number is n. (Thus if n is the Gödel number of $\phi(\sigma)$, then $t(n)$ is the Gödel number of $\phi(\bar{n})$.) We call $t(x)$ the *translation function*.

We call (T) a *Gödelian* theory if for every set A representable in (T), there is a *normal* Gödel sentence for A.

THEOREM 2. If there is a chameleonic Gödel sentence for $t^{-1}(A)$, then there is a normal Gödel sentence for A. More specifically, if $\phi(\sigma)$ is a chameleonic Gödel sentence for $t^{-1}(A)$, then its translation $\phi(\bar{n})$ is a (normal) Gödel sentence for A.

Proof. Suppose $\phi(\sigma)$ is a chameleonic Gödel sentence for $t^{-1}(A)$; let $\phi(\bar{n})$ be its translation. Since $\phi(\sigma)$ is a chameleonic Gödel sentence for $t^{-1}(A)$, then $\phi(\sigma) \in T^\sigma \leftrightarrow h \in t^{-1}(A)$. Also $h \in t^{-1}(A)$ if and only if $t(h) \in A$. Thus $\phi(\sigma) \in T^\sigma \leftrightarrow t(h) \in A$. Also $\phi(\sigma) \in T^\sigma \leftrightarrow \phi(\bar{h}) \in T$, if and only if $t(h) \in A$. But $t(h)$ is the Gödel number of $\phi(\bar{h})$. So $\phi(\bar{h})$ is a Gödel sentence for A.

COROLLARY 1. If $t^{-1}(A)$ is representable in (T), then there is a normal Gödel sentence for A.

Proof. Suppose $t^{-1}(A)$ is representable in (T). Then by Theorem 1 there is a chameleonic Gödel sentence $\phi(\sigma)$ for $t^{-1}(A)$. Then by Theorem 2, the translation of $\phi(\sigma)$ is a normal Gödel sentence for A.

From Corollary 2 we immediately get

THEOREM 3. A sufficient condition for (T) to be Gödelian is that the translation function $t(x)$ be admissible.

9. *Subtheories.* We now consider a theory (T) and a subtheory (P) of (T). We let T be the set of valid sentences of (T) and P be the set of valid sentences of (P). (We are thinking mainly of a complete consistent theory (T) and an axiomatizable subtheory (P), such as Peano Arithmetic. Then the elements of T are the *true* sentences and the elements of P are the *provable* sentences.) We continue to let T^σ be the set of chameleonic sentences whose translation is in T, and we let P^σ be the set of chameleonic sentences whose translation is in P. (If (P) is axiomatizable, and presented in the usual way by axioms and inference rules, then we can set up an axiomatic presentation of $(P)^\sigma$ as follows: We take as axioms of $(P)^\sigma$ the axioms of (P). Then we take as inference rules of $(P)^\sigma$ all inference rules of (P) *but restricted to normal sentences!* Then we add the additional inference rule: For any chameleonic sentence C, we may infer C from its translation. We then have an axiomatic system whose provable sentences are precisely those chameleonic sentences whose translations are in (P).)

THEOREM 4. (After Gödel) If $\overline{g(P^\sigma)}$ is representable in (T), then there is a normal sentence G which is in T but not in P.

Proof. Suppose $\overline{g(P^\sigma)}$ is representable in (T). Then by Theorem 1 there is a chameleonic sentence C such that $C \in T^\sigma$ if and only if $g(C) \in \overline{g(P^\sigma)}$. But $g(C) \in \overline{g(P^\sigma)}$ if and only if $C \notin P^\sigma$. Hence $C \in T^\sigma$ if and only if $C \notin P^\sigma$. Since $P \subseteq T$, then $P^\sigma \subseteq T^\sigma$, and so $C \in T^\sigma$ but $C \notin P^\sigma$. Therefore the translation G of C is in T but not in P.

Remark. The above could alternatively be obtained as a consequence of Corollary 1 of Theorem 1 as follows: The set $\overline{g(P^\sigma)}$ is the same as the set $t^{-1}\overline{g(P)}$, hence if $\overline{g(P^\sigma)}$ is representable in (T), then there is a normal Gödel sentence G for $\overline{g(P)}$, and so $G \in T \leftrightarrow G \notin P$. Since $P \subseteq T$, $G \in T$ and $G \notin P$.

COROLLARY 1. If we add to the hypothesis of Theorem 4 that (T) is consistent, then the theory (P) contains an undecidable sentence – i.e., a sentence G such that neither it nor its negation is in P.

COROLLARY 2. If (T) is a complete and consistent theory, (P) a subtheory, and if $g(P^\sigma)$ is representable in (T), then (P) contains an undecidable sentence.

COROLLARY 3. If (T) is a complete and consistent theory in which all recursively enumerable (r.e.) sets are representable, and if (P) is an *axiomatizable* subtheory of (T), and if the Gödel numbering g of the sentences of (T^σ) is recursive, then (P) contains an undecidable sentence.

Re Corollary 3, we must recall that the sets P^σ, T^σ are defined only relative to a Gödel numbering g (because the translations function is defined only relative to a given Gödel numbering). If g is recursive, so is the translation function, and since $g(P)$ is r.e. (by the hypothesis that (P) is axiomatizable), then $g(P^\sigma)$ is also r.e., hence representable in (T), and result follows by Corollary 2.

10. *Cross Reference.* We shall call (T) *doubly Gödelian* if for any representable sets A_1, A_2 there are normal sentences S_1, S_2 with Gödel numbers g_1, g_2, such that $S_1 \in T$ if and only if $g_2 \in A_2$, and S_2 is true if and only if $g_1 \in A_1$.

It is already known that Peano Arithmetic is doubly Gödelian. (This can be obtained from the double recursion theorem in much the same manner as the Gödelian property can be obtained from Kleene's Recursion Theorem. Also it can be obtained from Boolos's Generalized Diagonal lemma [1].) And the complete theory (N) of plus and times is doubly Gödelian – indeed, as we shall see, any theory in which all recursive functions are admissible must be doubly Gödelian (providing, of course, that the Gödel numbering is effective). However, the theorem that follows gives a somewhat stronger result, which we now wish to discuss.

Call two normal sentences X, Y *equivalent* (in the theory (T)) if they are either both in T or neither in T. (We do not require that the sentence $X \equiv Y$ be in T; this is a stronger equivalence relation which will not concern us here.) For any normal formula $\phi(x)$ and any sentence S, by $\phi[S]$ we mean the sentence $\phi(\bar{n})$, where n is the Gödel number of S. (An alternative notation for "$\phi[S]$" is "$\phi(^{\ulcorner}S^{\urcorner})$".) Now, to say that (T) is Gödelian is to say that for any normal formula $\phi(x)$ there is a normal sentence S such that S is equivalent to $\phi[S]$. And to say that (T) is doubly Gödelian is to say that for any two normal formulas $\phi_1(x)$, $\phi_2(x)$, there are normal sentences S_1, S_2 such that S_1 is equivalent to $\phi_2[S_2]$ and S_2 is equivalent to $\phi_1[S_1]$.

We now define (T) to be *strongly doubly Gödelian* if for any two normal formulas $\phi_1(x)$, $\phi_2(x)$ there are normal sentences S_1, S_2 such that S_1 is equivalent to $\phi_2[S_2]$ and S_2 *is* the sentence $\phi_1[S_1]$ (which of course implies that S_2 is equivalent to $\phi_1[S_1]$). In other words, (T) is strongly doubly Gödelian if and only if for any two normal formulas $\phi_1(x)$, $\phi_2(x)$ there is a normal sentence S such that S is equivalent to $\phi_1[\phi_2[S]]$. Stated in still another equivalent form, (T) is doubly Gödelian if given any set A representable in (T) and any normal formula $\phi(x)$ (but not necessarily one which represents A), there is a normal sentence S which is in T if and only if the Gödel number of $\phi[S]$ is in A. We are now interested in sufficient conditions for (T) to be strongly double Gödelian. What now follows occurred to us as a result of formalizing the construction of §4 of Part 1.

Consider a (not necessarily) normal formula $\phi(x)$ in which "x" is the only free variable. By its *substitution function* we mean the function $k(x)$ such that for all n, $k(n)$ is the Gödel number of $\phi(\bar{n})$. And we call a function $k(x)$ a *substitution function* if it is the *substitution function* of some formula $\phi(x)$.

THEOREM 5. If (T) is Gödelian and if all substitution functions for normal formulas are admissible in (T), then (T) is strongly doubly Gödelian.

Proof. Assume hypothesis. Take any representable set A. Let $\phi(x)$ be any normal formula, and let $k(x)$ be the substitution function for $\phi(x)$. By hypothesis, $k(x)$ is admissible, hence $k^{-1}(A)$ is representable. Since, by hypothesis, (T) is Gödelian, then there is a normal sentence S with Gödel number s such that $S \in T \leftrightarrow s \in k^{-1}(A)$. Now $s \in k^{-1}(A)$ if and only if $k(s) \in A$, but $k(s)$ is the Gödel number of $\phi(\bar{s})$. Therefore $S \in T$ if and only if the Gödel number of $\phi(\bar{s})$ is in A. This means that (t) is strongly double Gödelian.

As an immediate consequence of Theorem 5 and Theorem 3, we have the following:

THEOREM 6. If all substitution functions for normal formulas are admissible in (T), and if the translation function $t(x)$ is admissible in (T), then (T) is strongly doubly Gödelian.

Remarks. for any $n > 2$, define (T) to be n-fold Gödelian if for any normal formulas $\phi_1(x), \ldots, \phi_n(x)$ there are normal sentences $S_1, \ldots, S_n$ such that S_1 is equivalent to $\phi_2[S_2], \ldots, S_{n-1}$ is equivalent to $\phi_n[S_n]$, and S_n is equivalent to $\phi_1[S_1]$. Call (T) *strongly* n-fold Gödelian if for any normal formulas $\phi_1(x), \ldots, \phi_n(x)$ there are normal sentences $S_1, \ldots, S_n$ such that S_1 is equivalent to

$$\phi_2[S_2], \quad S_2 = \phi_3[S_3], \ldots, S_{n-1} = \phi_n[S_n] \text{ and } S_n = \phi_1[S_1].$$

It is not difficult to show that the hypothesis of Theorem 6 (or even the hypothesis of Theorem 5) implies that for every $n > 2$, the theory (T) is strongly n-fold Gödelian.

10.1. We would like to mention that Theorem 6 holds under a somewhat weaker hypothesis.

For any two functions f, g (from natural numbers to natural numbers) by the composite functions fg is meant the function which assigns to every n the number $f(g(n))$. It is obvious that if g, f are both admissible, then fg is admissible (because for any set A, $(fg)^{-1}(A) = g^{-1}(f^{-1}(A))$.

We leave the proof of the next two results as exercises for the reader.

LEMMA. For any function f, if the function ft is admissible (where t is the translation function) then for any representable set A there is a Gödel sentence for $f^{-1}(A)$.

THEOREM 6*. Suppose that for every substitution function $k(x)$ for normal formulas the composite function kt is admissible. Then (T) is strongly doubly Gödelian (and also strongly n-fold Gödelian for every $n \geq 2$).

11. *Simultaneous Self-Reference.* A (normal) formula $K(x, y)$ (with "x", "y" as the only free variables) is said to represent a binary relation $R(x,y)$ (on the natural numbers) if for all m, n the sentence $K(\bar{m}, \bar{n}) \in T$ if and only if $R(m,n)$, And $R(x,y)$ is called representable in (T) if it is represented by some normal formula $K(x,y)$.

Let us refer to the property of being Gödelian as property G_1. We shall say that a theory (T) has property G_2 if for any two binary relations $R_1(x, y)$, $R_2(x, y)$ which are representable in (T) there are normal sentences X_1, X_2 with respective Gödel numbers g_1, g_2 such that $X_1 \in T$ if and only if $R_1(g_1, g_2)$, and $X_2 \in T$ if and only if $R_2(g_1, g_2)$. (It is trivial to show that G_2 implies that (T) is doubly Gödelian; we do not know whether G_2 necessarily implies that (T) is strongly doubly Gödelian.) That Peano Arithmetic has property G_2 is a special case of the generalized diagonal lemma of Boolos [1]. (If we formalized the theory of two chameleonic symbols to arithmetic, the construction of §5 of Part 1 would go over just about intact to Boolos's proof.) We shall, however, approach property G_2 in a very different manner – one we derived from the construction of §6 of Part 1. This will yield an interesting property G_2^* which implies both G_2 and G_1. We first need the following definition.

What we have so far termed admissible shall now (for emphasis) be termed *admissible with respect to sets*: We shall say that a function f is admissible with respect to binary relations if for every representable relation $R(x, y)$, the relation $R(x, f(y))$ is also representable. (This incidentally implies that for every representable relation $R(x, y)$ the relation $R(f(x), y)$ is representable, because the inverse of a representable binary relation is again representable. It also implies that f is admissible with respect to sets, because for any representable set A, the set of ordered pairs $\langle x, y \rangle$ such that $y \in A$ is a representable relation.)

We shall henceforth use the term "admissible" to mean admissible with respect to binary relations (and bear in mind that this also implies admissibility with respect to sets).

Now we say that (T) has property G_2^* if for every representable relation $R(x, y)$, there is an admissible function $h(x)$ such that for every n, $h(n)$ is the Gödel number of a normal sentence, and for every n, $S_{h(n)} \in T$ if and only if $R(n, h(n))$.

LEMMA. Property G_2^* implies property G_1.

Proof. Suppose (T) has property G_2^*. Take any set A representable in T. Let R be the set of all ordered pairs $\langle x, y \rangle$ such that $y \in A$. Then R is a representable binary relation, hence by hypothesis there is a function $h(x)$ (an admissible one, in fact, but this is not relevant) such that for all n, $S_{h(n)}$ is normal and for all n, $S_{h(n)} \in T \leftrightarrow R(n, h(n))$. But $R(n, h(n)) \leftrightarrow h(n) \in A$. And so $S_{h(n)} \in T \leftrightarrow h(n) \in A$, which means that for *every* n, $S_{h(n)}$ is a Gödel sentence for A.

THEOREM 7. Property G_2^* implies property G_2.

Proof. Suppose (T) has property G_2^*. Take any two relations $R_1(x, y)$, $R_2(x, y)$ representable in (T). By property G_2^* (taking $R_2(x, y)$ for $R(x, y)$) there is an admissible function $h(x)$ such that for all n, $S_{h(n)}$ is a normal sentence, and $S_{h(n)} \in T$ if and only if $R_2(n, h(n))$. Since $h(x)$ is admissible, then the relation $R_1(x, h(y))$ is representable, and hence the set A of all n such that $R_1(n, h(n))$ is a representable set (because the relation $R_1(x, h(y))$ is represented by some normal formula $K(x, y)$, and the formula $K(x, x)$ then represents A). By lemma, (T) is Gödelian, hence there is a normal Gödel sentence S_a for the set A. Thus $S_a \in T \leftrightarrow R_1(a, h(a))$. But since for *any* n, $S_{h(n)} \in T \leftrightarrow R_2(n, h(n))$, then $S_{h(a)} \in T \leftrightarrow R_2(a, h(a))$. We let $b = h(a)$, and we see that $S_a \in T \leftrightarrow R_1(a, b)$ and $S_b \in T \leftrightarrow R_2(a, b)$. Thus we have property G_2.

We next seek a useful sufficient condition for a theory (T) to have property G_2^* Well, it turns out to be virtually the same as the hypothesis of Theorem 6.

For any normal formula $K(x, y)$ (with no free variables other than "x" and "y"), by $K(x, \sigma)$ we mean the result of substituting "σ" for every free occurrence of "y" in $K(x, y)$. ($K(x, \sigma)$ is thus a chameleonic formula with x as the only free variable.) We note that if $k(x)$ is the substitution function for $K(x, \sigma)$ (as previously defined) then for every n, $k(n)$ is the Gödel number of the chameleonic sentence $K(\bar{n}, \sigma)$.

THEOREM 8. If all substitution functions are admissible and if the translation function $t(x)$ is admissible, then (T) has properties G_2^*.

Proof. Assume hypothesis. Take any representable relation $R(x, y)$. Then the relation $R(x, t(y))$ is representable (because $t(x)$ is admissible). Let $K(x, y)$ be a normal formula which represents the relation $R(x, t(y))$. Consider the chameleonic formula $K(x, \sigma)$, and let $k(x)$ be its substitution function. By hypothesis, $k(x)$ is admissible. For any n, $k(n)$ is the Gödel number of $K(\bar{n}, \sigma)$, and therefore tkn is the Gödel number of $K(\bar{n}, \overline{kn})$ (which is the translation of $K(\bar{n}, \sigma)$). And so for all n, S_{tkn} is the Gödel number of the normal sentence $K(\bar{n}, \overline{kn})$. Also $K(\bar{n}, \overline{kn})$ is in T if and only if $R(n, tkn)$ holds (since $K(x, y)$ represents the relation $R(x, t(y))$). And so for every n, $S_{tkn} \in T$ if and only if $R(n, tkn)$. We let h be the composite function tk. Since t, k are both admissible, then h is admissible, and for every n, $S_{h(n)}$ is normal, and

$S_{h(n)} \in T$ if and only if $R(n, h(n))$. Thus (T) has property $G_2{}^*$.

From Theorems 7, 8 we immediately have the following:

THEOREM 9. If all substitution functions are admissible in (T) and the translation function is admissible in (T), then (T) has property G_2.

Remarks. Theorems 7, 8 and 9 are all related to the last construction of Part 1. If we worked in an arithmetic theory using class abstracts, the relation would be clearer. We hope to say more about this in a future publication.

12. *Normalizability.* We now turn to a problem mentioned in the introduction. Consider a chameleonic formula $K(x, \sigma)$ with x as the only free variable. For any set A, we shall say that $K(x, \sigma)$ represents A in $(T)^\sigma$ if for every n, the chameleonic sentence $K(\bar{n}, \sigma)$ is in T^σ if and only if $n \in A$. Now, if a set A is representable in $(T)^\sigma$, is it necessarily representable in (T)?

We shall put the matter another way: Let us say that a normal formula $\phi(x)$ *normalizes* $K(x, \sigma)$ if for every n, $\phi(\bar{n}) \in T$ if and only if $K(\bar{n}, \sigma) \in T^\sigma$. And we shall say that $K(x, \sigma)$ is *normalizable* if there is a normal formula which normalizes it. And so our question can be equivalently stated thus: Is every σ-formula $K(x, \sigma)$ necessarily normalizable?

In general, the answer would appear to be "no" (though we have not produced a counterexample). The problem is this: Of course $K(x, \sigma)$ has a translation $K(x, \bar{h})$ (where h is the Gödel number of $K(x, \sigma)$), but there is no reason to believe this translation normalizes $K(x, \sigma)$ (because, roughly speaking, every time we substitute a numeral $\bar{n}$ for "x" in $K(x, \sigma)$ – every time we take a particular instance $K(\bar{n}, \sigma)$ – the chameleonic symbol "σ" changes its denotation. Speaking more precisely, just because $K(x, \bar{h})$ is the translation of $K(x, \sigma)$, does not mean that $K(\bar{n}, \bar{h})$ is the translation of $K(\bar{n}, \sigma)$!).

Now, if all recursive functions are admissible in (T) (and if g is an effective Gödel numbering), then every chameleonic formula $K(x, \sigma)$ is indeed normalizable. More generally, we have the following result:

THEOREM 10. If all substitution functions are admissible in (T), then all σ-formulas with one free variable are normalizable. More specifically, given any formula $K(x, \sigma)$, if its substitution function $k(x)$ is admissible, then $K(x, \sigma)$ is normalizable.

Proof. Suppose the substitution function $k(x)$ of $K(x, \sigma)$ is admissible. The (normal) formula $K(x, y)$ represents some relation $R(x, y)$ in (T). Since k is admissible, then the relation $R(x, k(y))$ is representable in (T). Hence the set A of all n such that $R(n, k(n))$ is representable in (T); let $\phi(x)$ be a normal formula which represents A. We will show that $\phi(x)$ normalizes $K(x, \sigma)$.

For any n, $\phi(\bar{n}) \in T \leftrightarrow R(n, k(n))$ holds. And $R(n, k(n))$ holds if and only if the (normal) sentence $K(\bar{n}, \overline{k(n)}) \in T$. But since $k(n)$ is the Gödel number of $K(\bar{n}, \sigma)$, then $K(\bar{n}, \overline{kn})$ is the translation of $K(\bar{n}, \sigma)$. Therefore $K(\bar{n}, \overline{kn}) \in T \leftrightarrow K(\bar{n}, \sigma) \in T^{\sigma}$. And so $\phi(\bar{n}) \in T \leftrightarrow K(\bar{n}, \sigma) \in T^{\sigma}$, which means that $\phi(x)$ normalizes $K(x, \sigma)$.

13. *Concluding remarks.* (1) Adopting a neat idea of Tarski [7], we can prove the following stronger form of Theorem 1: For any set A representable in (T), there is a chameleonic Gödel sentence C which contains only one occurrence of "σ". To prove this, suppose $\phi(x)$ is a normal formula which represents A. Then, instead of taking the chameleonic sentence $\phi(\sigma)$, we take the chameleonic sentence $\exists x(x = \sigma \wedge \phi(x))$.

Indeed, we could have gotten by using only chameleonic formulas which contain at most one occurrence of "σ". The translation function would then be easier to arithmetize (that is, the task of showing it to be admissible for a particular theory (T) would be easier), since it would involve the operation of replacement (substitution for one occurrence) instead of the more complex operation of substitution for an unknown number of occurrences of "σ".

(2) Somewhat related to the above, we would like to give a very brief sketch of an alternative type of chameleonic arithmetic language.

Starting with a given theory (T), instead of adjoining an individual chameleonic symbol "σ", we could adjoin a predicate H which we call the *chameleonic predicate*. (Roughly speaking, in any context, $H(x)$ means the same as $x = \sigma$ – viz., that x is the Gödel number of the whole context.) More precisely, we define a chameleonic formula as a formula built from H and the original predicates and function symbols of (T). In fact we need consider only those chameleonic formulas C in which H has at most one occurrence, and this occurrence is always followed by the same variable – say "x". Then we define the translation of C to be the result of replacing Hx by $x = \bar{c}$, where c is the Gödel number of C. Then we define T^H as the set of all C whose translation is in T.

The set T^H of course plays the role of T^σ. (For example, to prove Theorem 1 for T^H rather than T^σ, given a normal formula $\phi(x)$ representing a set A, the sentence $(\exists x)(Hx \wedge \phi(x))$ is the chameleonic Gödel sentence for A – that is, it is in T^H if and only if its Gödel number is in A.)

This plan is particularly suitable if we are working with a *pure* arithmetic theory (T) – that is, one in which there are no numerals to name numbers, but instead, there are formulas $Z_0(x)$, $Z_1(x)$, ..., $Z_n(x)$, ... such that for each n, $Z_n(x)$ is satisfied by n and by no other number. Only in working with such theories, we would have to redefine the translation of C to be the result of replacing Hx by $Z_c(x)$, where c is the Gödel number of C.

(3) Going back to our chameleonic languages of the σ-type, it is not really necessary to adjoin a new symbol "σ" to the formalism; we could choose one particular variable v and call it the *chameleon symbol* (and frame all relevant definitions accordingly). In this case, the translation function wouldn't be very different from Gödel's diagonal function (the only difference is that the diagonal function involves substitution for all *free* occurrences of v, whereas the translation function involves substituting for *all* occurrences of v). What would then have to play the role of the "chameleonic formulas" would be those formulas involving no free variables other then v and in which v does not occur as a bound variable.

We chose to add the new symbol "σ" mainly for heuristic purposes; without it, the notation of "chameleonic truth" would be an unnatural one, and surely the idea of normalizability would never have occurred to us.

(4) If we were working in a first-order theory with class abstracts (like the arithmetic language treated in [3]), then we would define a chameleonic sentence to be of the form $H\sigma$, where H is a class abstract (but not one which contains σ). Then we would define the translation of $H\sigma$ to be $H\bar{n}$, where n is the Gödel number of $H\sigma$. The resulting translation function would be simpler yet (though a shade less simple than the norm function of [3]).

(5) The notion which we have introduced of a function f being *admissible* has been basic in just about all our proofs. We believe this to be a useful notion for first-order theories in general (quite apart from chameleonic systems). We should mention some related notions.

In some theories, there are primitive recursive terms for all primitive recursive functions. Now, if there is a term $t(x)$ for a function $f(x)$, then of course $f(x)$ is admissible in our sense (if $\phi(x)$ is a formula which represents A, $\phi(t(x))$ will represent $f^{-1}(A)$).

For first-order theories with identity, Tarski [6] defines a function $f(x)$ to be *definable* if there is a formula $K(x, y)$ such that for all m, n the sentence $\forall y(K(\bar{n}, y) \equiv (y = \overline{f(n)})$ is in T. If a function $f(x)$ is definable in this sense, then it is admissible.

There is the weaker notion of the relation $f(x) = y$ being *completely representable* in (T) – which means that there is a formula $K(x, y)$ such that for all m, n, $f(m) = n$ if and only if $K(\bar{m}, \bar{n}) \in T$, and $f(m) \notin n$ if and only if the negation of $K(\bar{m}, \bar{n}) \in T$. For an ω-consistent theory, this is enough to guarantee that $f(x)$ is admissible.

Finally, there is the still weaker notion of the relation $f(x) = y$ being *representable* in (T). For *complete* consistent theories, this is enough to guarantee admissibility.

(6) It might be noted that except for the brief remarks under (2) none of our proofs have made any use of the logical connectives or quantifiers, and yet we have been talking about first-order theories! Roughly speaking, this is because our notion of *admissibility* – unlike the related notion of definability in the sense of Tarski, which involves the existential quantifier and the biconditional – does not involve logical connectives or quantifiers. It should therefore not be surprising that all the results we have proven for first-order theories go over to the more general setting of representation systems [5]. (The notion of a function being admissible goes over intact). And there is no problem in defining an appropriate notion of a *chameleonic* representation system. But this is a subject for another article.

NOTE

[1] In *The Devil's Dictionary*, the author Ambrose Bierce, after defining the word "I", writes: "Its plural is said to be *We*, but how there can be more than one myself is doubtless clearer to the grammarians than it is to the author of this incomparable dictionary."

REFERENCES

[1] Boolos, George: 1979, *The Unprovability of Consistency*, Cambridge University Press, Cambridge.

[2] Gödel, Kurt: 1931, 'Über formal unentscheidbar Sätze der Principa Mathematica und verwandter Systeme I' *Monatshefte für Mathematik und Physik* **38**, 173–198.

[3] Smullyan, Raymond M.: 1957, 'Languages in Which Self Reference Is Possible', *Journal of Symbolic Logic* **22**, 55–67.

[4] Smullyan, Raymond M.: 1978, *What Is the Name of This Book?* Prentice-Hall, Inc. Englewood Cliffs.

[5] Smullyan, Raymond M.: 1959, *Theory of Formal Systems*, Annals of Mathematics Studies #47, Princeton University Press, Princeton.

[6] Tarski, Alfred: 1953, *Undecidable Theories*, North-Holland, Amsterdam.

Dept. of Philosophy
City University Graduate Center
33 West 42nd St.,
New York, NY 10036
U.S.A.

RAPHAEL STERN

RELATIONAL MODEL SYSTEMS: THE CRAFT OF LOGIC

PART 1

This is an essay in the foundations of language; it is not solely an essay in logic.[1] It is concerned *both* with conceptual and technical matters; it is not, then, intended simply as a technical exercise. We must, however, given our conceptual leanings here, adopt a somewhat different technical stance; hence we must also exhibit a technical device and begin to discuss its computational properties (but we only begin to do so, we are *not* concerned here to give completeness proofs, etc.).

For quite some time now the technical devices used to study natural languages have seemed somewhat inept. In worrying about how to formally reconstruct sublanguages of a natural language and in attempting to mend fences a bit, we have, I warrant, opened a Pandora's box with issues of all sorts tumbling out, a veritable mine of issues and worries, including issues about foundations of language, about mathematics, about logic, worries about how to theorize, about rationality, metaphor, and about how to explicate epistemic concepts.

After some thought I came to the conclusion that there might be a remedy for some of our problems if we could but see our way to enriching the theories we were using to study language, and I thought this task might be helped along if, at least in part, the technical devices so important to these theories *and* embedded in the theories, were themselves enriched. Looking about there seemed two places at which one could enrich our technical devices: we might alter the set theory upon which many devices depend, or one might enrich one or more of the technical devices dependent upon the set theory. I thought it best to do the latter. This essay is about one such effort at enrichment. (I don't for one moment mean to suggest that the two tasks are mutually exclusive – one could enrich both. However, if one does introduce enough enrichment into the technical devices dependent upon the set theory, there turns out to be a decided disinclination to *also* enrich the latter as well.)

This enrichment is one part of the task of revising our views about

Synthese **60** (1984) 225–252.   0039–7857/84/0602–0225 $02.80
© 1984 *by D. Reidel Publishing Company*

how best to deal with natural languages. I thought I would introduce the device here and begin the task of describing it and then rather briefly, and only at the end of this essay, compare it to another system (McArthur's) to see whether the two systems tend to look equivalent from a computational point of view. In this essay I go a little way toward showing that they are computationally equivalent. It turns out that given two devices, D and D', one can maintain computational equivalence between the two while enriching D considerably (and not also enriching D'). What we need to decide is what is the most suitable set of enrichments preserving computational equivalency that best accords with still other needs, namely, to get still other jobs done. Depending upon the size of the theory in which D is embedded, and the number of tasks the theory addresses, we may find ourselves introducing very rich devices indeed.

The material covered here is motivated by four major concerns and a host of minor ones: (i) we wish to make an effort to get a better understanding of what is needed for an account of rational acceptance; (ii) we wish to see just what is involved in theory construction and here we focus on problems connected with the technical and nontechnical requirements of theorizing over certain domains; (iii) we are, of course, concerned about natural languages; and finally (iv) we feel we ought to pay some attention to views about metaphor and especially to views concerning the proper role of metaphor in theorizing, particularly since metaphor has, for a long while now, been held to evidence pretheoretical thinking and since many thinkers still hold that metaphor is best dispensed with once real straightforward theorizing begins. These issues are quite definitely related but it is too much to expect to go very deeply into all of the relations here. Perhaps, though, to motivate and illuminate some of our concerns in this essay, we should focus on one dilemma because it is central to our interests and quite revealing once we are concerned to provide accounts of theorizing, of natural languages, and of rational acceptance.

Suppose we cite some current thinking about metaphor. Many people hold one or more of the following: (i) Some thinkers claim that metaphor belongs to and is appropriate to our poetic uses of language and that we should aim, on the whole, to make our discourse more scientific; (ii) others claim that metaphors "carry" ideologies that prejudice theorizing; and (iii) still others claim that metaphorical uses of language are pretheoretical, the mark of an absence of theory, and so for them it seems to follow that a mark of proper theorizing is that we have finally dispensed with metaphor altogether in our theorizing.

With this much on their plate, theorists then worry about *how* to get rid of metaphor. Abstract, formal devices are a favorite ploy. What better route, some think, than to approach theory construction by aiming at a degree of formalization and abstraction that rids one of metaphor (this creates a false, worrisome and confusing opposition between technique and metaphor, also between theory and metaphor; also, it does, in part, account for the rather rapid growth in the use of formal devices in various fields for it provides some ideological justification for a rather unrelenting deployment of these devices hither and yon). What better, some think, than to use the set-theoretic apparatus to achieve this end and to invest the set-theoretic tradition with these aims in particular since the set-theoretic tradition addresses theory, for we can utilize the formal techniques and the reductive tendencies there, or so the argument would run, to invest theory with the technical devices needed to weed out metaphor. Thus, the set-theoretic tradition becomes an arbiter of an ideology, becomes invested with an ideology, indeed, becomes the standard bearer in this crusade against metaphor.

In this essay I take a rather different approach. The essay, in part, works out the consequences of a different group of assumptions about rationality, about theory, about metaphor, about simplicity as an aesthetic guide for proper theorizing, about enrichments and formal devices. I hold, for example, that it is not, on the whole, the task of theory, and hence of the set-theoretic tradition, to get rid of metaphor for us. Nor is it a sign of proper theorizing to hew to a methodology which takes account of metaphor only to the extent of dispensing with it; if one does do this one loses out on proper models for a good deal of our theorizing, one is likely to get theories which do not resonate with other theories, and one is likely to encounter difficulties in getting an account of rational acceptance off the ground. On the contrary, here we hold that theory, wherever possible, is supposed to help us to *retain*, and select metaphors appropriate to a particular theoretical enterprise *and* to a certain domain, and then theory refines our metaphors a bit by reconstructing them in the light of their technical counterparts (which the theory also enables us to introduce). The role of set theory on this view is to enable us to introduce the technical counterparts to metaphors (in the formal part of our theory) and then also to reconstruct the metaphors, with the slightly reconstructed metaphors (reconstructed in the light of their formal counterparts) abiding at the lower levels of our theory. Metaphorical uses of language can be obstructive, difficult, intrusive and vague, but it is the task of theory to relieve us of this onerous aspect of metaphor without throwing the conceptual baby out with the bathwater. If this is so then we need to be wary of that aspect of the set-theoretic tradition, what I tend to think of as its *ideological core*, that has made set theory the handmaiden of certain views about theory. The logical devices that have grown up in the set-theoretic tradition, have as a result of this ideological core, taken one form rather than another and as a result have – given the handiwork of logically minded, philosophically inclined antimetaphorists – been denuded of many symbolic virtues.

Instead, here, we wish to provide a technical device that has graph theory as one of its underpinnings and that uses the graph-theoretic apparatus to resonate with commonplace metaphors about *time*, with epistemic concepts as well, and that enables us to reconstruct metaphors relevant to the domains of several subtheories of our larger theory, and thereby that provides greater unity over our theories and greater resonance between subtheories of our larger theory.

A clue to the sorts of jobs we need done would also come from a brief look at conditions for an account of rationality. (We can in virtue of this "look" begin to see the ties between rationality and the role of metaphor.) There are three concepts centrally involved in providing accounts of rational behavior: (i) whether someone has looked to or sought *alternatives*; (ii) the nature of an individual's *anticipation* of future results; and (iii) the way an individual uses *information*. These three concepts – information, anticipation, and alternative – are all at the service of accounts of rational behavior, where accounts of rational behavior must be competent to specify appropriate *fits* among beliefs, or appropriate *fits* among actions, or among theories. These accounts must also specify acceptable departures from standard fits and consequently what counts as an acceptable deformation in a space of beliefs, or actions, or theories. And embedded in our talk about anticipation and about finding alternatives, are metaphors about the future and the past that are crucial to include and develop if we are to get a handle on the way we deal with *fits* and *misfits* among actions, beliefs and theories. And central to reconstructing, technically, these metaphors is the notion of "taking a futurewalk", with the relevant metaphors reconstructed by using a technical development, the concept of a 'futurewalk'.

What I aim to do, then, is get a computationally adequate device founded in graph theory, that also enables us to get a grip on concepts like anticipation, and in designing this system I do so with respect to another system, McArthur's, making it richer than his but computationally equivalent to it.

In the course of this I suggest how to revise our views about metaphor, about theory, about how to use formal devices, etc., and I do this in the hope that we might dispense with the ideological core of the set-theoretic tradition, or at least the onerous part of it, and so that as a result we are no longer quite so disinclined to countenance metaphor. Our views on how to theorize and on how to use set theory and related theories have in the past been too closely tied to our unfortunate views about metaphor and its purported relations to proper theorizing.

PART 2

I should like to construct the valuational component of a tense logic. The question is just how we shall go about doing this. There are various ways to proceed. For example, one could, following McArthur (1976), define a function which indexes truth-value functions and then define an ordering, R_{mc} (I rename McArthur's binary relation, "R_{mc}"), on the function indexing the truth-value functions. There are a number of reasons why I do not wish to proceed in precisely this way. I would prefer to construct a relation similar to R_{mc} which I will call "R'" (I introduce this later on) and then by constructing a set J of graphs, also introduced later on, define a valuation function V. It turns out that this way of proceeding, combined with a logical device which I model after Hintikka model sets, and which I call "relational models", substantially enriches our semantical theory.

What to expect from this essay: This is merely a sketch of a device. We provide the essential elements to understand how the device is to work, including a step-by-step comparison with McArthur's system for K_t, and then we show by considering a few cases how the two systems (McArthur's and mine) compare with regard to valuations (when McArthur's system assigns a wff X a T, for true, does our system assign a comparable value P, etc.?) and this too will begin to show how the two systems compare, and how this system works. We do not, here, include soundness or completeness proofs. We are interested now in establishing the technical groundwork for later work of this sort. This essay is, after all, an essay in the *conceptual* foundations for studies of language – it is not, per se, an essay in logic. The logical device is demanded by the changes in methodology and is presented as a way of discovering and following out the full implications of my changes in methodological principles. It seems to me that the device introduced should be taken in the same spirit as the remarks earlier about metaphor and as the remarks I make later about *crafts*: they are all important parts of this essay and all of them are dealt with in more detail elsewhere and all constitute parts of a work in progress. None of the remarks "work" *alone* but in this all-too-short a space there is little time to go into more detail than I already have done about any one of them. As far as I and others who have checked the material can tell, the devices do the work they are constructed to do – proofs would follow later. As far as other aspects of the technical constructions, I benefitted greatly from discussions with and suggestions from Wilfried Sieg, but I daresay here, too, there are alternatives one could pursue as well that I shall consider elsewhere (there are many ways, for example, to introduce the graph-paths, and one would have to look at all of the uses to which the theory is being put to decide on one or another; at the suggestion of Professor Sieg I have chosen one way and there is no question of it being "wrong" but there may very well be more efficacious routes).

Why enrich? This is a question that tends to worry people particularly in the light of the plethora of technical devices that visit our poor planet each year and the difficulties attending each new birth (completeness and soundness proofs, etc.). I gave some sense of this just now in discussing metaphor but perhaps a few additional words will not prove amiss. In developing a tense-theoretic system, indeed, in developing many another semantical device, we usually expect to use some logical device that is modelled in or developed from set theory. These devices will then, if past performance is any indication, rely quite heavily on set theory. So we expect to approach our task with a hierarchy of formal theories, with set theory at the pinnacle of the hierarchy, and whenever this obtains we really ought to ask ourselves whether this suffices or whether the domain under study requires something more, something there is a tendency nowadays to overlook, namely, whether the system is sufficiently rich. Along with Goodman in this volume, I tend to think that the typical uses of set theory, in particular when coupled with ideologies such as the one involving the role of theory vis-a-vis metaphor, tend to foster fewer models, hints, problem-solving suggestions, etc., rather than more – tend, then, to denude our other theoretical efforts of conceptual possibilities rather than enrich these efforts. (Thus, all of this and what follows both reenforce the difficulties already discussed with regard to theory and metaphor and disavow as readily the uses to which the formal apparatuses have been put in the rather dubious service of getting rid of metaphor.) In the form in which we currently pursue our set-theoretically founded studies we do not tend to foster enough interesting relations among different theories, in particular where these theories interface.

Where, you may ask, *do we enrich*? Rather than, as Goodman does, seeking an enrichment by enriching the set theory itself, we tend, here, to enrich at other places in the hierarchy of theories which "stand" on top of the set theory. We enrich in two places: by introducing a new, somewhat richer logical device called a "relational model", pursued without Occamist constraints, and by altering the valuation function, in this case the function V, and defining it with respect to a set *J* of graphs.

Why, you may insist, *do we bother with enrichments*? Suppose I make some attempt to spell this out in more detail. What we aim to do is the following: we want to find a nondenumerable family of sets, the relational-model system, and we want to develop a system so as to maximize the number of relations used in the system. We also want to be able to build up the sets in the system by putting wffs into the sets. We then look for rules of construction which will enable us to put wffs into the sets and then, depending upon our needs, we will define fewer or more relations over the sets. Our aim is to develop the richest system possible consistent with our theoretical needs. In doing this we abandon the "Occam razor"–style approach that logicians typically have used when they built systems of this sort, and we abandon the aesthetic constraints that have motivated them. What they have attempted to do – that is, what the standard methodology calls for and has called for for quite some time now – is to build systems using the fewest relations, entities and rules. Under the guidance of the aesthetics of an Occam razor–style methodology, simplicity is the keynote. But this has depleted our symbol systems to the point where they now resonate poorly with other theoretical constructions which depend upon these symbol systems. I drop the methodological constraints and, instead, try to build the richest systems possible so that we tend to take far larger classes of relations than are strictly needed to build our sets or to gain computational equivalency with other semantical devices. As it turns out mere computational equivalency, for the narrow range of computational tasks defined as interesting to logicians, no longer suffices once the logical devices are put to use in larger theoretical contexts.

What, then, is our aim? The aim of the logical devices which we introduce here is a *symbolic* one as well as a technical one, and we want the devices to be rich enough to satisfy symbolic functions – namely, we want the devices to be rich enough to "point to" other classes of theories and their related research programs, and so that they can provide models for solutions to problems, and models for new theories, and so that they provide an enrichment and illumination over a fairly wide domain of other theories. (It goes without saying that in the course of this, decisions about the appropriate attitude to take to metaphor arise.)

One of Goodman's complaints (this volume) has to do with precisely this sort of issue: registered somewhat in my terms, Goodman does not think that set theory as currently pursued and used provides sufficient richness; it is lacking, then, in the symbolic dimensions mentioned above. He raises questions about set theory and its relation, for example, to physics. Set theory, he thinks, as currently pursued, that is *unenriched* set theory, tends to denude mathematics, when mathematics is reconstructed set theoretically, of the richness needed to properly resonate with physics; set-theoretically reconstructed mathematics is then no longer able, as mathematics, according to Goodman, once was, to answer to the needs of physicists and of mathematicians. I am inclined to agree with Goodman, on the whole. However, I tend to ascribe our difficulties both to the set-theoretic tradition *and* the apparatus; for it is certainly clear enough that something must be done both with the apparatuses we are inclined to construct and with the tradition. I find that what Goodman has to say about the relations of mathematics and physics are equally true of other uses of set theory and so, also, of relations of mathematics and natural languages.

A bit of historical information might not go amiss here. As I already indicated, things were not always this way. Indeed they are not quite this way now, for there is even now a bit of a tug of war between mathematicians who have one ideal of abstraction and of degrees of formalization that a problem requires and some very logically minded logicians. Nonetheless, things did go from bad to worse around about 1910. But it is sometimes difficult to account for the degree of exasperation that some of the early analytic philosophers felt and expressed at frontal and hind assaults from formalisms. I think to some degree at least they felt that the devices simply did not work all that well for their own purposes. It was, however, a bit difficult to understand the insistence with which each camp purveyed its own goods (more of this later), but each showed a craftsmanlike intensity of purpose. However, one could not *give up* formal systems insofar as both camps agreed, and so they did, that when studying language we are evincing sets of structures that are preferred by speakers of a language – these structures could only be studied with formal systems. Nonetheless the task did seem clear enough – we would have to do something about the devices. It always seemed to me that we did *not* need to revert back to informal, analytic analyses in order to cope, that we could by some method(s) of enrichment maximize the tasks our devices could address.

The solution, at least part of one, that seemed indicated for us was that when someone proposed a set of technical devices to use to analyze some area or other, we ought to look at them *and* at the theories used to reconstruct them, such as set theory, and we ought to test the following assumptions: (i) that *set theory* is needed to reconstruct the theories; (ii) that we need to reconstruct the theories; and (iii) that the systems, reconstructed, are rich enough. Thus, (i) is an assumption we frequently make but it is not essential to use set theory to reconstruct other theories; we could find some other theory to do the job. And (ii) is an assumption one could make and would tend to make were one operating within the set-theoretic tradition; but neither (ii) nor (iii) are "givens". Indeed, we could dispute anywhere along the line. I tend, for example, to hold for purposes of this essay that (i) is so, but that (iii) is not. I tend to hold elsewhere that (iii) is false over a very large number of theories and domains and that (i) is true but *only* for certain theories and not for others – there are at least two very basic theories one could use to reconstruct other theories and not simply the one.

We ought, then, to look about us rather carefully to see whether our devices are adequate and rich enough. And then, if they are not adequate or rich enough, we ought to seriously consider how and whether to enrich them, and if we do decide that some enrichment is needed, we need, also, to worry about whether to enrich the set theory itself or some other theory (or theories) in the hierarchy of theories that includes set theory. And finally, if we do enrich some theory or other, we need to decide on some class of devices to use to do so. And so with regard to many theoretical tasks that commit us to formal devices, we are, I warrant, committing ourselves (when we use such devices) to a host of assumptions, as above, any one of which we can question.

Here we disagree with Goodman about where the enrichment should take place, not about whether enrichment is needed. So in what follows we shall propose a certain sort of enrichment, not of the set theory itself, but of the logical devices we use. And to do the latter we introduce another device, one with a number of relations (used in essential ways in defining the device), and the relations suffice as "points of attack", "focal points" suitable for enrichment. In every case, what we do is bring to bear on the construction of the relation one or another mathematical device, usually a standard device (such as graph theory or topology or automata theory), and we use *this* device to construct the relation differently – we proceed uniformly over all enrichments by "attacking" the relations and constructing them in unexpected ways. But for safety's sake, we tag some other system, such as McArthur's, for which completeness and soundness results have been developed, and we construct our devices in parallel with one or another of these (such as McArthur's) in this essay, so that we achieve computational equivalence with respect to standard, logical-type computations. The constraint, then, in introducing new devices, is that we are to be conservative over changes in a certain class of computations, namely, a class of computations defined by the "usual" logics, and then we "radicalize" the systems at other points, allowing for enrichment.

If one thinks of *doing logic* with an Occamist ideology – that is, of introducing devices and systems with the fewest rules, entities, and relations – then we are not doing logic here but something very much like logic. You might, with some justice, suggest that on this view we are doing mathematics, but we are doing mathematics under a particular set of constraints, namely, that whatever we produce when we construct our devices (these mathematical ones) they must at least obey certain constraints, and these are constraints introduced with regard to standard ways of doing logic – the systems, for example, must preserve standard logical-type computations. If, however, you are willing to abandon the Occamist ideology, then you may think of this, too, as doing logic.

We should now, with all of this, get on to the business of specifying the logical device. In doing this we need the following: rules of construction in order to develop the family of sets, relations and a set of graphs, and finally a valuation function. We also will need a step-by-step comparison between this and McArthur's system to see just how well the systems compare with regard to certain computations.

PART 3

Logic: *General*. In constructing a logic, a number of decisions are needed: we want to know just what our representational theories will be like (logics can be thought of as, at times, reconstructing or *re*-presenting, formally, concepts, operations, etc., that appear on other levels of our theory), we want to test the assumptions guiding our construction of these devices, see just how "thin" or rich the devices will be, and we wish to see just how realist-oriented or nominalist-oriented we are going to be, and then we take matters from there. In this case I have introduced a theory of relational models that can be viewed as intermediate between Hintikka model sets and Hiż's truth sets, or, even better, as a rewriting of the theory of model sets which takes that account in a rather different direction. My aim, here, is to construct systems that are as rich as possible.

The need for formal analyses of a host of objects, in a great variety of areas, requires the diversity I am recommending. For example, if a natural language is your interest, then you are stuck with a system in which an unexpectedly large number of formal devices find models. This creates difficulties: your device might find a model in the system but only in some trivial sublanguage which, for lack of a realistic approach to the language, you view, instead, as a significant and very large part of the natural language; your device may provide insight and reconstruction of an important subset of the computations and structures of interest with regard to the natural language but be singularly unrevealing in explaining why things are as they are. It is rather extraordinary that a natural language should be quite so rich, but there you are, and to accomodate this degree of richness we do need devices which are sufficiently rich as well. To achieve this end we introduce this "new" technical device rather than sticking to model sets.

Logic: *Relational Model Systems*

I wish to provide a semantic analysis of one tense-theoretic system, Lemmon's minimal system K_t. The analysis will use relational model systems and it is nominalistically oriented. Apart from our purely technical interest in just how this would be done, analyses of tense notions are important for studies in the philosophy of science, for accounts of rationality, and for studies of subsystems of English. Our formal devices presuppose a set theory (ZF) *without* enrichment of the set theory itself; however, the formal devices are themselves enriched. These enrichments anticipate the grab bag of uses to which these devices will be put. To develop a tense logic we shall need a family T of sets of wffs and relations specified on the sets in T and on T itself.

Our formal device is a relational model system over a formal language L. The effect of this is to introduce a class of preferred structures on L. L consists of all of the wffs of the propositional calculus (abbreviated "PC") *and* wffs of the form "HX", "GX", "FX", and "PX". The relational model system T is a family of sets, with T infinitely large. Ordinarily, T, or something like T, would enable one to introduce the maximal class of preferred structures on L that are consistent with the smallest number of assumptions, rules, relations and so on. We drop this last constraint. (Not the consistency part, of course.) Each Tm (here "Tm" stands for a set, or relational model in T, the relational model system) in T contains infinitely many wffs of L. Every Tm in T is a proper subset of L. Each Tm but one contains atomic wffs: there is one and only one Tm in T that contains no atomic wffs, and one and only one Tm in T that contains all of the atomic wffs of PC. There are at least two sorts of relations involved when we use relational model systems: relations on each of the sets in T, and relations on T itself. A three-place predicate "... is as true as ... in ..."(a *wff* is as true as another *wff* in *a relational model*) will suffice for the relation on the sets in T but we are going to need one (or more) relations on T itself. The indexing relation $\mathscr{L}$ is one relation on T; so is the relation R', a dyadic relation on $\mathscr{L}$. Instead of just using a dyadic relation I have decided to also introduce a set J of graphs, tied to R', which will be used to define the valuation function V. J lends itself to 'travel' *and* a host of other metaphors, is constructed with the help of graph theory, is a set of ordered pairs with each pair consisting of a set of relational models plus a set of ordered triples formed on the index set I (I is an index set and the ordered triples are all of the form $\langle n, n+1, n+2 \rangle$), and based as it is on graph theory, J is consonant with our talk about *time* and *anticipation*, and with the intuitions, ideas, and styles of thinking infecting our discourse and ideas about time.

Axioms for the three-place relation '... is as true as ... in ...', abbreviated "R'''", are as follows:

(i) X is as true as X in Tm;
(ii)   If X is as true as Y in Tm then Y is as true as X in Tm;
(iii) If X is as true as Y in Tm and if Y is as true as Z in Tm then X is as true as Z in Tm.

With this and with I- and R-rules, we will have the wherewithal to define T, the family of relational models. Insofar as specifically tense-theoretic matters are concerned, we need to define a set J of graphs, with J and R' being used to define V the valuation function. R' constrains J – what is in R' as an ordered pair of indexed relational models $\langle Tm, Tm' \rangle$ must appear in J as nodes of at least one graph; and nodes of graphs are all and only the pairs in R' – nothing else can serve as a node. But there is a whole set of temporal theories. We study here the minimal system K_t, but there are other more complex theories. As we shift from K_t to still other theories, and insofar as these are more complex, we add features to the relation, R_{mc} and also to R'. When this happens, the construction of J also changes. The sorts of graphs we put into J change with different theories, and so we really have a set of sets, with J as one of them, and yet also with *other* sets of graphs like J but different, each reflecting the changing complexity of the theory and the relations we use. Also, for each theory such as K_t, and for each fixed construal of R' (one and only one construal per tense-theoretic theory), we can construct J in a variety of different ways, and so for each construal of R', we can have several different ways of constructing J. This may reflect different uses to which we put our overall theory, or slightly different decisions about which metaphors to retain when we wish to talk about *time, anticipation, language*, and so on.

The relational models can be thought of either as worlds or as cross-sections of a single world, or as temporal states, etc. We will use "P" for the past-tense operator, "F" for the future-tense operator, "G" will read "it will always be the case that," and "H" will read "it has always been the case that." In defining a semantics for these operators we shall need to define a *moment*.

I omit in this essay any reference to proof theory for I intend to introduce it elsewhere. Instead, I shall concentrate on the first stage of our development of the semantic apparatus, the specification of the *I*- and *R*-rules. This pair of rules enables us to define any relational model in the family T of sets of relational models. In doing this we use a three-place relation R''. R'' enables us to devise rules of construction for the relational models, and we can view the rules as specifying constraints on what does and what does not go into a *Tm*. And when some wff X is in *Tm*, then we say that X is preferred, or that it is preferred *in* or *on* or *relative to Tm*. The sets of rules appear below; we let "AT" stand for the set of all atomic wffs of PC.

I-rules

$$(X_1, X_2, Tm) \in R'' \text{ iff } X_1 \in Tm \ \& \ X_2 \in Tm.$$

R-rules

1. There are infinitely many wffs in each of the sets in T.
2. T itself is nondenumerably large.
3. One and only one relational model, called "a-Tm", contains no atomic wffs.
4. We define "NEG(AT)" as the set of all X's such that $X = -Y$ and such that Y is an atomic wff.
5. NEG(AT) $\subset a$-Tm.
6. There is one and only one relational model, b-Tm, such that AT $\subset b$-Tm.
7. No two *Tms* contain the same atomic wffs.
8. An (denumerably) infinite number of *Tms* contain but one atomic wff in each of them; an infinite number of *Tms* contain but two atomic wffs in each of them; an infinite number of *Tms* contain three and only three atomic wffs in each of them; an infinite number of *Tms* contain four and only four atomic wffs in each of them; and so on for five, six, etc., atomic wffs.[2]
9. Either $(X, X, Tm) \in R''$ or $(-X, -X, Tm) \in R''$, but $(X, -X, Tm)$ is never an element of R''.
10. $(X_1, X_2, Tm) \in R''$ iff $(X_1, X_1 \ \& \ X_2, Tm) \in R''$.

What does all of this mean? To get a glimpse of how this works we might point out that these sets operate in very much the same way that Hintikka model sets operate. To get some clearer notion perhaps a comparison to truth-value functions would do. If we consider how one sets up a truth-value function, we can see that when we do so we admit a mapping from the set of atomic wffs to (T, F), and we can see that there are nondenumerably many such mappings. On one such mapping, *all* atomic wffs are mapped to T, on another *none* are mapped to T, and perhaps on an infinite number of other mappings just one atomic wff is mapped to T while all the others are mapped to F. Now for each truth-value assignment g that maps a wff, say "p", to a truth value such as T, there is a corresponding relational model Tm to which for example, p belongs and so which does the same sort of valuational job with regard to P as g does with regard to T, so that when p gets T on g then p gets P on the Tm in virtue of being a member of the Tm. What we are getting at is that for each truth-value assignment g, there is a unique Tm performing an equivalent valuational task. We understand, of course, that Tms do not assign T, that membership in a Tm enables us to allot a value P to wffs instead of T, that this assignment is relative to a Tm, and that when we allot P we say things like p gets P on Tm, or p gets P with respect to Tm. And here the value P plays very much the same role that T does. When we assign P, however, we mean that a wff which is P, is preferred *relative to* or *on* a Tm to which it belongs. (Notice two things: (i) there are two sorts of computations involved here: one is assigning values and might be dubbed the "logical" one, and the other is afforded by graph theory; (ii) wherever possible we begin to substitute "prefer" for "true" to avoid the metaphysical difficulties introduced by *truth*, denoting semantics, and so on, and the "bad press" *truth* gets in the literature.)

These rules determine the relational models, and so, also, they determine the family of sets, T. Once we determine T we can then give it a bit more structure, and we can define for this purpose a function $\mathscr{L}$. After defining $\mathscr{L}$ we then go on to provide still more structure by defining a dyadic relation R', and then J, and with all of this done we can then define a *moment* and then also the valuation function V. Since J represents a departure from more standard approaches to tense-theoretic constructions, and since it requires graph theory, we shall need to explain what this is all about. We go into this in the section labelled "valuation". There we enrich our device by founding it in some fashion in graph theory. The difficulty was to decide just where to enrich; I find that it is plausible to do so when we introduce V.

Indices, Graphs, Walks and Futurewalks

We want an example of how an enriched semantics would work. Rather than enrich our set theory, I choose to enrich the hierarchy of theories that stem from and depend upon the set theory by introducing a graph-theoretically based account of tense notions. Let me explain just how this works.

A tense-theoretic system has certain requirements. We need to distinguish states of a world viewed, perhaps as an ongoing process or system, or, if you will, with a different metaphysics at the root, we might think of our need as one of requiring a distinction between different worlds, and so on. Once we are clear about this aspect of our theory, we then decide on a representational theory to *re*-present our somewhat vague intuition. This is frequently set theory, and we then look to find a family of sets to do proxy for and *re*-present the notion of a *set of worlds* or of a *set of states*. Once we decide how to construct the sets, then we need to decide, also, just how much more structure we are going to give the sets. We proceed here as follows: first by introducing rules of construction, also R' and $\mathcal{L}$, we provide some structure on the family of sets. Then we proceed to a second stage and we provide a bit more structure by introducing J and V. We have already defined the family of sets T. The next bit of structure we now add involves the index relation on the set T and the index set I. I consists of real numbers and we define $\mathcal{L}$ by pairing each Tm in T with at least one i in I. Next we impose a bit more structure by defining R' in $\mathcal{L}$. R' is dyadic, is defined on indexed relational models, and we allow R' to be a set of pairs of (indexed) relational models and we allow R' to have any combination of properties. (R' is the precise analogue of McArthur's relation.)

In doing all of this we don't subscribe to Occamist conventions – we *don't* define the smallest class of relations, sets, rules, etc., consistent with getting the computations done. Rather we intend to define a far richer class of structures, richer than what we would need to get computational equivalency with McArthur's system. We do so to enrich the technical apparatus and to interpose graph theory between logic and set theory so that rather than interpreting logic directly in set theory, we first interpret graph theory in set theory and then interpret our technical devices in graph-theoretic terms. Defining a moment is rather simple: A moment is an ordered triple $\langle \mathscr{L}, R', T_m \rangle$. In order to define V we need one more item, the set J of graphs. We now proceed to introduce the notion of a graph, just a bit, and discuss J. After this we can define the valuation function V.

The Set J of Graphs

We shall need a set J of graphs. In order to understand the constitution of J we should say a few words about graphs and about the constraints on J.

We shall set things out sequentially as numbered points.

(i) Every j in J is a graph.

(ii) A graph has nodes (or *points* or *vertices*) and lines.

(iii) For our purposes here the points (or nodes or vertices) of a graph are *Tms* (relational models).

(iv) We take a line here to be an unordered triple of elements of I.

(v) A graph is a finite set of points together with a finite set of unordered triples (unordered triples of items out of I).

(a) To each line there corresponds two points.

(b) A directed graph is a finite set of points together with a set of ordered triples of elements of I.

(vi) Examples of graphs would be the following:

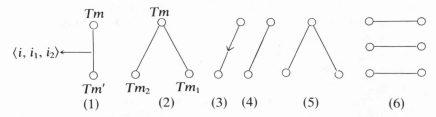

$\langle i, i_1, i_2 \rangle \leftarrow$

Tm Tm

Tm' Tm_2 Tm_1

(1)   (2)   (3)   (4)   (5)   (6)

(I put up these different graphs to provide some idea of the sheer variety of graphs. Thus, (3) is directed, and (4) is not. Also, (6) is a larger graph consisting of graphs. In the case of (2) we labelled the nodes as *Tms*; in the case of (1) we labelled both the nodes and the line.)

(vii) For convenience we can also define an element. This is a line, (i, i_1, i_2), precisely as in (1) above, along with two vertices, also as in (1) above, one on each end of the line. The line is called the *line of the element*.

(viii) A graph G may also be defined as a set of elements; the elements are said to be the elements of G or the elements belonging to G; if e' is an element of a graph G, then the dot(s) of e' can be thought of as a dot of G, and also as a dot of e'; also the line of e' can be thought of either as a line of e' or a line of G.

(ix) We say that an element is incident to a dot or vertex and that a dot or vertex is incident to an element when the dot is the vertex of the element; we also say that a line is incident to a vertex if the line is the line of an element e' and the vertex is incident to the element e'.

(x) The degree of a vertex, V in or belonging to a graph G is the number of elements of G incident to V.

(xi) In the context of a theory for K_t we are interested only in graphs, two of whose vertices are of degree 1, the rest of degree 2 or more. Since we can define J in a number of ways even for K_t, we can select one particular way of construing J here. For our purposes, then, J consists of graphs such that for each graph G, two and only two vertices of G are of degree 1, the rest of degree 2.

(xii) *Start* and *end* vertices are vertices of degree one.

(xiii) The length of a graph G is the number of lines in G.

(xiv) For graphs G of length greater than 1, a dot is a start of G if it is of degree 1 in G, and if it is incident to an element, the line of which is a triple, where the sum of the components of the line is lower than the sum of the components of the triple that constitutes the line of the element incident to the other dot of degree 1. The *end* element is the dot of degree 1 that is not the start element.

(xv) A t consists of atomic wffs of the form $p_1, p_2, \ldots$. We define the numerically smaller atomic wff of any pair to be that atomic wff with the lower subscript.

(xvi) For graphs of length l, the start node must have at least one atomic wff that is numerically smaller than any of the atomic wffs in the end node.

(xvii) A subgraph of a graph is a finite set of elements of the graph.

(xviii) A walk of a directed graph G is a subgraph G' of G, with G' a sequence in which points and lines alternate and each line is incident to the points immediately preceding and following it.

(xix) A path of a graph is a walk of a graph only two of whose vertices are of degree one with all of the other vertices of degree 2 but no higher.

If G' is a subgraph of G and is either a path or a walk, we say that G' is a path, or we say that G' is a walk, and if Tm is the start node and Tm' the end node of G', then we also say that there is a walk G' from Tm to Tm', or, there is a walk from Tm to Tm', or G' is a walk from Tm to Tm'.

(xx) We then have in J directed graphs whose nodes are Tms. For this particular account of K_t, and for any account, and usually relative to one's aims and needs, one can choose one rather than another construal of J. Each graph G in J obeys certain constraints on node choice. Thus if (X, Y) are nodes, X a start and Y an end of a subgraph G' of G, then $(X, Y) \in R'$. Moreover, if (X, Y) belongs to R' then there exists a directed graph G with subgraph G', such that X is a start node of G' and Y is an end node of G'. For example, refering to our figures, in figure (1), suppose (1) were a directed graph and Tm the start, then (Tm, Tm') would be in R'.

(xxi) A futurewalk G' from Tm to Tm' is a subgraph G' of a graph G, such that G' is a directed graph, and G' is a walk from Tm to Tm'. In such cases we also say that G' is a futurewalk, or that there is a futurewalk G'. We acknowledge that a graph G' is a trivial subgraph of a graph G, when $G' = G$. We allow that G' is a *pastwalk* from Tm' to Tm when G' is a futurewalk from Tm to Tm'.

(xxii) We shall discuss this in more detail later but it should be evident by now that there is a one-to-one, onto map from truth-value functions to Tms. We call this function "$\mathscr{L}'$". Given the constraints on J, and that we call here McArthur's binary relation "R_{mc}", later discussion will establish that when $\langle f, g \rangle$ belongs to R_{mc}, for f and g truth-value functions, then $\langle \mathscr{L}'(f), \mathscr{L}'(g) \rangle$ belongs to R', and then, also, there exists in J a graph G such that there is a subgraph G' of G and G' is a futurewalk from a node $\mathscr{L}'(f)$ to a node $\mathscr{L}'(g)$.

The Valuation Function V

We now have the wherewithal to construct the valuation function. It will turn out that V is very rich and yet computationally adequate; it is also the case that relative to one theory K_t and one choice of the relation R', we can have several different constructions of the set of graphs, hence, several different constructions of the valuation function. This is but one. Each function maps wffs to P (for preferred) or to $\not{P}$ for not preferred. The pair $(P, \not{P})$ plays the same sort of role that (T, F) and $(1, 0)$ play.

For "A" over atomic wffs and "X", "X_1", "X_2", etc., over any wffs at all:

1. $V(A, (\mathscr{L}, R', Tm)) = P$ when $(A, A, Tm) \in R''$, hence when $A \in Tm$; otherwise $V(A, (\mathscr{L}, R', Tm)) = \not{P}$.

2. $V(-X, (\mathscr{L}, R', Tm)) = P$ when $(-X, -X, Tm) \in R''$, hence when $-X \in Tm$; otherwise $V(-X, (\mathscr{L}, R', Tm)) = \not{P}$.

3. $V(X_1 \& X_2, (\mathscr{L}, R', Tm)) = P$ when both $(X_1, X_1, Tm) \in R''$ and $(X_2, X_2, Tm) \in R''$, hence when both X_1 and also $X_2 \in Tm$; otherwise $V(X_1 \& X_2, (\mathscr{L}, R', Tm)) = \not{P}$.

4. $V(FX, (\mathscr{L}, R', Tm)) = P$ when there exists a moment $(\mathscr{L}, R', Tm')$ such that there exists a graph, $G, G \in J$, and there is a subgraph G' of G such that G' is a futurewalk from a start node Tm to an end node Tm' and $V(X, (\mathscr{L}, R', Tm')) = P$; otherwise $V(FX, (\mathscr{L}, R', Tm)) = \not{P}$.

5. $V(PX, (\mathscr{L}, R', Tm)) = P$ when there exists a moment $(\mathscr{L}, R', Tm')$ such that there is a graph G of J, and a subgraph G' of G with G' a futurewalk from Tm' to Tm and where $V(X, (\mathscr{L}, R', Tm')) = P$; otherwise, $\not{P}$.

6. $V(GX, (\mathscr{L}, R', Tm_i)) = P$ when, for every moment $(\mathscr{L}, R', Tm_j)$ such that there exists a graph G of J with G' a subgraph of G and G' a futurewalk from Tm_i to Tm_j, $V(X, (\mathscr{L}, R', Tm_j)) = P$; otherwise, $\not{P}$.

7. $V(HX, (\mathscr{L}, R', Tm_i)) = P$ when for every moment $(\mathscr{L}, R', Tm_j)$ such that there exists a graph G of J with G' a subgraph of G and G' a futurewalk from Tm_j to Tm_i, $V(X, (\mathscr{L}, R', Tm_j)) = P$; otherwise, $\not{P}$.

A Comparison of Truth-Value Assignments and Relational Models

We should like to compare these systems to begin to feel a bit reassured that they are computationally equivalent. To begin our comparisons of the two systems we ought first to compare the construction of truth-value assignments with the construction of Tms in some depth. Suppose we think of "AT" as standing for the set of all atomic wffs. A truth-value assignment g is a map of each element W of AT to either T or F (McArthur uses $(0, 1)$; one of the few departures we make here in describing his system is that we use (T, F) whereas he uses $(0, 1)$). Each truth-value function g is a set of pairs. Each truth-value function is denumerably infinite and the set AT is also denumerably infinite. In one truth-value assignment g there will be ordered pairs (W_1, T), (W_2, T), (W_3, F), and so on. There are nondenumerably many truth-value assignments in McArthur's system.

Suppose we now consider just one such assignment g with just one and only one atomic wff p mapped to T under g. We then also can say that "p is T on g". We assume that all of the remaining atomic wffs are paired with F under g. Besides g there are also a denumerably infinite number of other possible combinations if we are interested in all ways of pairing just one atomic wff in AT with T; we might for example, pair p_1 with T under g_1 and then pair all the rest of the wffs in AT including p with F, or we might, instead, under still another function, pair p_2 with T and pair all the rest with F, and so on for still other functions and other methods of pairing wffs with T and F.

Now given this, we construct a Tm which has just the atomic wffs *in it* that are paired with T under g. When we designed the Tms we constructed them according to the following sort of constraints:

$$\text{For } W \in \text{AT}, \; W \in Tm \text{ iff } \langle W, T \rangle \in g$$
$$\text{For } W \in \text{AT}, \; W \notin Tm \text{ iff } \langle W, F \rangle \in g.$$

This means that every time we pair an atomic wff with T we put that wff into some Tm, and every time we don't pair it with T but with F we do *not* put it into that Tm. With regard to the truth-value assignment g all of the atomic wffs that get T on g are paired with T and so get into Tm, and the rest that are not paired with T but with F do *not* get put into Tm. Thus we are constructing Tms with but one atomic wff in them, some with but two atomic wffs in them, some with but three, and so on. Now, just as with Hintikka model sets or with Hiż's truth sets where wffs in the model sets are thought of as true, so here wffs that are in a Tm are thought of as preferred, or as P, with respect to that Tm, or can be thought of as P *on* a Tm. We use the pair of values $(P, \not\!P)$ where McArthur and others use $(1, 0)$ or (T, F). If a wff, W, is in Tm then it is P, or it gets the value P on the Tm.

A Tm is constructed, as in this case, so that just those atomic wffs getting T on some function like g, are in the Tm and so get P, and so we find there is an exact correspondence in the two systems between assigning T on some function like g and assigning P on some relational model like Tm. Ultimately we are aiming for the following more general relation. For "W" over atomic wffs:

W is true on g_i iff W gets T on g_i
W is true on g_i iff W gets P on Tm_i
$W \in Tm_i$ iff $\langle W, T \rangle \in g_i$,

and these three equivalences hold for any $i \in I$, with I an index set indexing the truth-value functions and the relational models. And we understand all of this as indicating that for each function g_i, there is one and only one relational model corresponding to it, namely, Tm_i, so that each does comparable valuation work or value-assignment work with regard to atomic wffs.

Suppose, to illustrate just what we are driving at, we consider still other truth-value assignments that assign T to just one and only one atomic wff. These are $g^1, g^2, g^3, \ldots$. For each we pick a different atomic wff and pair it with T. Thus, for g^1 we have p_1 and pair it with T, and for g^2 we have p_2 and we pair it with T, and so on. In the case of each g^i all of the atomic wffs not paired with T are paired with F. And we can, as before, construct relational models corresponding to each one of the truth-value assignments and the corresponding Tms would be as follows: $Tm^1, Tm^2, Tm^3, \ldots$. Clearly Tm^1 corresponds to g^1 and so contains only p_1, and Tm^2 corresponds to g^2 and so contains only p_2, and so on. The same would hold for all of the other g^i and the Tm^i.

As before we can consider still other cases. We can consider truth-value assignments that assign T to two and only two atomic wffs, with the remaining atomic wffs assigned F. Here is a list of those functions assigning T to two and only two atomic wffs per function:

$g^1_1, g^2_1, g^3_1, \ldots$.

Suppose we consider just one case, g^1_1. We assume that

(p_1, T) and (p_2, T) both belong to g^1_1.

If this is so then there exists a Tm such that both p_1 and p_2 and only these two atomic wffs belong to that Tm. Suppose this is Tm^1_1. We then suppose that corresponding to each of the truth-value assignments there are Tms of the following sort:

$Tm^1_1, Tm^2_1, Tm^3_1, \ldots$.

And we suppose that there is an exact and one-to-one correspondence between the truth-value assignments and the Tms.

Now all of this would be true for truth-value assignments that assign only three atomic wffs a T, and also for those that assign only four atomic wffs a T, and for those that assign 5, 6, 7, and so on, a T. We can always find a corresponding Tm that performs comparable work with regard to P that the truth-value assignments perform with regard to T. There is, then a one-to-one, onto function, $\mathscr{L}'$, mapping the truth-value assignments to the Tms. We have, then the following relations:

for "W" over atomic wffs and "M" over Tms

W is true on $\mathscr{2}$ iff W is P on $\mathscr{L}'(\mathscr{2})$

$\langle W, T \rangle \in \mathscr{2}$ iff $W \in \mathscr{L}'(\mathscr{2})$

$\langle W, F \rangle \in \mathscr{2}$ iff $W \notin \mathscr{L}'(\mathscr{2})$.

What of other aspects of the two systems? In the two systems we index the family of sets. And so, for $i \in I$ whenever McArthur assigns an index i to some truth-value assignment $\mathscr{2}$, we assign i to $\mathscr{L}'(\mathscr{2})$, and *vice versa*.

McArthur then goes on to specify what he means by a *moment*. He does so by specifying a relation over indexed truth-value sets. He understands a truth-value function to be a temporal state of the world and we understand the corresponding Tm to also be a temporal state of the world. Where McArthur imposes a dyadic relation on indexed truth-value assignments, we impose ours on indexed Tms. For any moment $\langle \Omega, R_{mc}, \mathscr{2} \rangle$ in McArthur's system we can construct one and only one moment in my system, and we do so by substituting $\mathscr{L}$ for Ω, and then substituting R' for R_{mc} and finally $\mathscr{L}'(\mathscr{2})$ for $\mathscr{2}$.

We call McArthur's binary relation "R_{ms}". R_{mc} is allowed to have any combination of properties and is defined over the indexed truth-value assignments; we introduce a binary relation R', over the corresponding indexed relational models and we allow R' to have any combination of properties as well.

Thus, when for McArthur, $\mathscr{2}$ is R_{mc} to $\mathscr{2}'$, for us

$\mathscr{L}'(\mathscr{2})$ is R' to $\mathscr{L}'(\mathscr{2}')$, and also we find that $\mathscr{L}'(\mathscr{2})$ is on a walk to $\mathscr{L}'(\mathscr{2}')$ and also that there is a futurewalk from $\mathscr{L}'(\mathscr{2})$ to $\mathscr{L}'(\mathscr{2}')$.

Computational Equivalence

We have built a device comparable to that of McArthur's but one that is richer. McArthur has shown that his system is complete. If we can show that whenever his system assigns a wff X a T we assign it a P, and when we assign X a P, his system assigns it a T, then we are very close to seeing both that the systems are computationally equivalent and that this system is complete as well. We are not going to prove completeness here; we shall build up confidence in the system and show how it works by setting out a few cases and comparing the two systems in one direction and seeing whether when McArthur's system assigns the wffs in our test cases a T, we assign it P. To facilitate things let me set out a list of equivalences:

(i) X is T on $\mathscr{2}$ iff X is P on $\mathscr{L}'(\mathscr{2})$
(ii) X is F on $\mathscr{2}$ iff X is $\not{P}$ on $\mathscr{L}'(\mathscr{2})$
(iii) X is T on $\mathscr{2}$ iff $\langle X, T\rangle \in \mathscr{2}$
(iv) X is F on $\mathscr{2}$ iff $\langle X, F\rangle \in \mathscr{2}$
(v) X is P on $\mathscr{L}'(\mathscr{2})$ iff $X \in \mathscr{L}'(\mathscr{2})$
(vi) X is $\not{P}$ on $\mathscr{L}'(\mathscr{2})$ iff $X \notin \mathscr{L}'(\mathscr{2})$
(vii)   For the index relations it holds that $\langle \mathscr{2}, i\rangle$ iff $\langle \mathscr{L}'(\mathscr{2}), i\rangle$
(viii) $\langle \mathscr{2}, \Lambda\rangle \in R_{mc}$ iff $\langle \mathscr{L}'(\mathscr{2}), \mathscr{L}'(\Lambda)\rangle \in R'$ iff there exists a graph G in J and a subgraph G' of G and G' is a futurewalk from $\mathscr{L}'(\mathscr{2})$ to $\mathscr{L}'(\Lambda)$

With this we can get on to a few cases.

CASE 1. A is atomic.

We suppose a truth-value assignment $\mathscr{2}$ that assigns T to A so that A is true on $\mathscr{2}$ at moment $\langle \Omega, R_{mc}, \mathscr{2}\rangle$. We want to show that there exists a comparable moment in our system and that our system assigns P to A when McArthur's assigns it T.

The comparable moment is easy to find. We substitute $\mathscr{L}$ for Ω and R' for R_{mc} and $\mathscr{L}'(\mathscr{2})$ for $\mathscr{2}$. The question now is, Does A get P at $\langle \mathscr{L}, R', \mathscr{L}'(\mathscr{2})\rangle$? Suppose, as we are given, A gets T on $\mathscr{2}$. Then $\langle A, T\rangle \in \mathscr{2}$. But then by our equivalence, $A \in \mathscr{L}'(\mathscr{2})$. But then $\langle A, A, \mathscr{L}'(\mathscr{2})\rangle \in R''$. But then $V(A, (\mathscr{L}, R', \mathscr{L}'(\mathscr{2})) = P$. But then V assigns a P to A when $\mathscr{2}$ assigns a T to A.

CASE 2. For $Y = (X \mathbin{\&} Z)$, $-Y$ is true on $\mathscr{2}$, for X, Z atomic.

We suppose $-Y$ is true on $\mathscr{2}$ at moment $(\Omega, R_{mc}, \mathscr{2})$.

If this is so then Y gets F on $\mathscr{2}$ and if this is so then either Z or X or both get F on $\mathscr{2}$.

The moment corresponding to McArthur's is easily found by substituting $\mathcal{L}$ for Ω, and R' for R_{mc} and $\mathcal{L}'(\mathcal{Q})$ for $\mathcal{Q}$. Given that either (Z, F) or (X, F) or both $\in \mathcal{Q}$ then by our equivalences either $Z \notin \mathcal{L}'(\mathcal{Q})$ or $X \notin \mathcal{L}'(\mathcal{Q})$ or neither belong to $\mathcal{L}'(\mathcal{Q})$. But then by our rules of construction #10 $(Z \& X) \notin \mathcal{L}'(\mathcal{Q})$, and then by rule #9, $-Y \in \mathcal{L}'(\mathcal{Q})$; but then $\langle -Y, -Y, \mathcal{L}'(\mathcal{Q}) \rangle \in \dot{R}''$. Now given the construction of V, V then assigns P to $-Y$. But then when $\mathcal{Q}$ assigns T to $-Y$, V assigns P to $-Y$.

CASE 3. FX is true and X is atomic.

It follows that on McArthur's system FX gets T at some moment (Ω, R_{mc}, g), when $R_{mc}(g, f)$ and X gets T at (Ω, R_{mc}, f).

First we find the two corresponding moments. These are $(\mathcal{L}, R', \mathcal{L}'(g))$ and $(\mathcal{L}, R', \mathcal{L}'(f))$.

We are given that $(g, f) \in R_{mc}$. By our equivalences it follows that $\langle \mathcal{L}'(g), \mathcal{L}'(f) \rangle \in R'$. But then there exists a moment, $(\mathcal{L}, R', \mathcal{L}'(f))$ and a futurewalk from $\mathcal{L}'(g)$ to $\mathcal{L}'(f)$ and it now behooves us to see if X is P at $\langle \mathcal{L}, R', \mathcal{L}'(f) \rangle$. But X gets T at (Ω, R_{mc}, f), and so $(X, T) \in f$ and by our equivalences $X \in \mathcal{L}'(f)$, but then $(X, X, \mathcal{L}'(f)) \in R''$, but then V assigns X a P at $(\mathcal{L}, R', \mathcal{L}'(f))$, but then V assigns FX a P at $(\mathcal{L}, R', \mathcal{L}'(g))$, but then when FX is true on g it gets P on $\mathcal{L}'(g)$.

I think these three cases ought to suffice to show the reader how this system works and how the two systems compare.

Metaphor and Enrichment

My views here about how to enrich a logical device are a function of the fact that I embed the devices in a hierarchy of theories and let the theories, in part, dictate the aims of the devices. The aim of enrichment is to maximize the symbolic and explanatory virtues of the device while preserving computational efficiency.

We have seen that to some degree Occamist tendencies with regard to theorizing have gone hand in hand with our views of metaphor. In changing my view of metaphor and so also the role of theory, I have also abandoned the Occamist constraints of theorists. Occamist ideals counsel simplicity; we abandon this because a lot of the motivation for this ideal stems from a certain view of the symbolic role of theories, a certain view of rationality and a certain view of metaphor. As all of these, in particular the last, change, so do our methodological constraints. Thus, according to a typical antimetaphorist, the presence of metaphor is supposed to indicate laziness and sloppiness in thinking, with the thinking not "trim"; metaphorical thinking is, or so goes the tale, overloaded with vagueness, misdirection, unneccessary baggage, bad semantic practices, and an improperly conceived, overendowed universe. Metaphor reflects, on this view, an inability to get at the basic structures of the domain one is studying; it reflects, as well, bad logic, a pretheoretical kind of thinking that is unable to cope with the complexities and realities underlying a domain. We shall call this the "Platonic" view of metaphor. On the Platonic view if we are to do what one might call proper theorizing, the best way to do things seems to be to refine ones formal apparatus – for example, one's set-theoretically controlled devices – to hone and refine them until we divest them of metaphor. The ideal then becomes a set theory and technical device that is as "thin" as possible, with the fewest assumptions and technical enrichments. This ideology, now incorporated into the set-theoretic tradition, places the onerous task of getting rid of metaphor on our technical devices and on our theories, and it literally weds the cry for simplicity to the need to get rid of metaphors; it marries Occamist designs *to* Platonism *to* our set-theoretically founded, technical devices and, as well, *to* our set-theoretic tradition, and to our theories, of course. This is not necessarily a wedding we want to countenance but one we seem forced into because of our views about metaphor, etc.

Here we take it that it is the task of theory not to get rid of metaphor but to constrain and control metaphor, and to find the most efficacious way of introducing it and reconstructing it. All of this is done by a hierarchically organized theory with the controls (the reconstructing, technical devices) at one level of the theory and the (properly controlled) metaphor at another level of the theory. We want our formal devices to resonate properly with one another, and we want theory to provide the needed links between metaphor and our formal devices.

For some time now antimetaphorists have coupled with their view of theory the view that the proper place for metaphor is poetry. But it is theory as well as poetry, and not poetry alone, that inculcates the proper use of metaphor over a domain – if we revert to poetry alone we get a very skewed view of metaphor.

Our task is to choose a system with an abundance of relations, for this is the crucial point in theory construction. When we find relations we are going to reconstruct them and intrude one or more mathematical theories at that juncture, where relations obtain, to provide reconstructions of ideas and metaphors, and to provide models and hints about problem-solving techniques. We create a system which maximizes these possibilities and this is a relational model system whose task it is to introduce a maximal number of relations relative to a set of tasks while maintaining computational equivalency.

Logicians have a fair amount of resistance to the idea of enrichment and to these rather different methodological principles. Practitioners of logic are today wholeheartedly committed to the set-theoretic tradition and to a degree of abstraction unheard of in earlier times. (One has but to read Cantor to espy the differences and to see how, just a few decades ago, formal matters were construed.) This is not always bad; it contributes to the high degree of professionalism one finds among logicians. There are, however, unrewarding features as well. This method of doing logic is but one of many methods, one of many stages in a long history and, however recalcitrant logicians might appear, this is merely one way of practicing logic, sanctified only by very recent tradition.

I think that along these lines (to understand the strength of the resistance, the conservativism, the teeth gnashing over new methods) it is best to view science and logic as crafts. This entails that many of the metalogical ideals such as that logical principles are eternal truths, or, that they are everywhere *true* are merely felicitous myths, quasi-religious ideals that one takes to one's bosom, just the sort of thing that craftsmen *are wont to say to strengthen their fiber*, so that they can push on and get the job done, so to speak. If we view logic-doing as a craft, then we demythologize logic, and we can then begin to admit that it makes sense to build onto logics this clutch of mathematical systems I espouse to enable us to get the tasks we need done under way. It is not strange to find that the attitude of the craftsman is that *his* principles are sacrosanct (until, of course, the next one comes along, indifference sets in about the last but one and he thoroughly espouses the new); he has always done a lot of historical handwaving about things like eternal principles so we are not inclined to take this bit of handwaving any more seriously than the last. Thus once we recognize that these are often the feints and attitudes of a craftsman, and nothing more, once these "truths" are viewed as principles adopted for certain purposes, then we are a step or two closer to revising ones own views and admitting with some grace that, often enough, one constructs a logical device within or with respect to a cluster of theories and that one can construct the device with a maximal number of points at which one can focus mathematical systems for enrichment – enrichment and the whole baggage of attendant mathematical systems are plausible once we view logic-making as a craft. Once viewed as a craft, we are that much closer to admitting that ideas such as that logical principles are eternal are not "truths," but merely commitments made by craftsmen, things they want to believe, perhaps even ideals to approximate in their work; for in constructing systems so that they approximate such ideals they expect to get the best results they know relative to their aims and to get these results for the largest class of systems. If we acknowledge our role as craftsmen, then we can begin to admit something like the following: these "truths" are in fact the best principles we have for a wide range of purposes and they maximize our results relative to a class of systems in which we are interested and likely to remain interested for quite some time to come.

The principles, Occamist and otherwise, that a craft espouses at some historical period define the tasks of the craftsman for some time, but they can be replaced by still other principles. This happens all the time. When we dignify the wrong principles as truths we are in danger of refusing change when older systems and principles no longer redress problems we wish answered.

NOTES

[1] I wish to express my thanks to Wilfried Sieg for his considerable help, patience, and suggestions.
[2] The point is that we "take out" pairs (triples, etc.) from AT in every possible combination and "distribute" them over the Tm_i. The same of course for triples, quadruples, etc., from AT.

REFERENCE

McArthur, R.: 1976, *Tense Logic*, D. Reidel, Dordrecht.

340 Riverside Drive
New York, NY 10025
U.S.A.

J. DILLER AND A. S. TROELSTRA

REALIZABILITY AND INTUITIONISTIC LOGIC

1. INTRODUCTION

1.1. The present paper is an attempt to clarify the relationship between Heyting's well-known explanation of the intuitionistic logical operators on the one hand, and realizability interpretations on the other hand, in particular in connection with the theory of types as developed by P. Martin-Löf. Part of the discussion may be regarded as a supplement to the discussion by Martin-Löf.

1.2. *Heyting's explanations* were formulated more or less in their present form in 1930, and first published in Heyting 1934. Kolmogorov (1932) published the interpretation[1] in terms of problems, which later came to be regarded by Heyting (1958) as essentially equivalent to his own explanations. For later statements, see e.g., Heyting 1956, 1974.

1.3. *Realizability* by means of partial recursive functions, for intuitionistic arithmetic, was introduced by Kleene (1945); later Kleene developed a parallel notion of realizability by functions (detailed description in Kleene-Vesley 1965; for information on notions of realizability, see also Troelstra 1973). Finally, an abstract version of realizability in an arbitrary applicative structure (from which Kleene's notions can be recovered) was formulated by Feferman (1975; corrected 1979, IV. 2).

The parallel between Heyting's explanations and Kleene's realizability did not pass unnoticed (e.g. Troelstra 1973, 3.2.1; van Dalen 1973, p. 26; Dummett 1977, pp. 318–329, Kreisel 1971, pp. 142–144, 159–163). However, as becomes clear from his retrospective survey (1973), Kleene himself was not inspired by Heyting's or Kolmogorov's explanations, but rather by the finitist explanation of existential statements as incomplete communications (Hilbert-Bernays 1934, p. 32).

1.4. *Formulae-as-types.* Curry, in Curry-Feys 1957, pp. 312–315 observed that there is a correspondence between the axioms for

Synthese **60** (1984) 253–282.   0039–7857/84/0602–0253 $03.00
© 1984 *by D. Reidel Publishing Company*

intuitionistic implicational logic and the types of the basic combinators S, K; this observation was the germ of the formulae-as-types idea ("propositions-as-types" would have been more accurate), which found a clear expression in Howard 1980 (in circulation as a preprint since 1969). Independently of Howard, and more or less simultaneously, the formulae-as-types idea is exploited in Scott 1970, Läuchli 1970 and de Bruijn 1970.

In the formulae-as-types approach, one thinks (roughly) of a proposition as given by the set[2] of its "proofs", so propositions are associated with special sets. In the case of implications and universal statements, these sets are sets of functions. Thus the meaning of implication and universal quantification depends on the concept of function used. In intuitionistic logic these proof-functions are taken to be constructive (= lawlike, or lawlike relative to certain choice-objects occurring in the statements considered).

The most consistent and radical use of the formulae-as-types idea is to be found in P. Martin-Löf's papers (1975, 1982; henceforth we shall quote these papers as ML 75, ML 82). The very elegant formalisms developed in these papers are, at least at first sight, rather different from the more conventional formalizations of constructive/intuitionistic logic and mathematics.

The systems of ML 75, ML 82 can be said to model (in an informal sense) Heyting's explanations. In its thorough-going extensionality ML 82 seems to be somewhat farther removed from Heyting's intentions than ML 75. In lectures and conversations Martin-Löf has also stressed the motivation for his formalisms as deriving from a theory of meaning (cf. also the discussion in Sundholm 1983). Of necessity, we have to be somewhat brief about this aspect, since we are primarily concerned with the connections with realizability here; a short sketch is given at the end of section 3.

1.5. *Plan of the paper.* We begin with a recapitulation of Heyting's explanations and give a brief resumé of Sundholm's proposal (1983) for resolving certain difficulties connected with Heyting's explanations (section 2).

Then we discuss the informal notions of "basic domain" and "basic equality on a basic domain"; these notions[3] cannot be grasped independently of each other. The formulae-as-types idea then provides the interpretation of logic in terms of these notions. Varying the requirements on "basic equality on a basic domain" thus influences the logic. The formulae-as-types idea is closely connected with Sundholm's proposal to distinguish between "construction-as-a-process" and "construction as result of a process". Informal proofs (arguments) are of the first kind; only the second meaning corresponds to *mathematical* objects.

The "proofs of a proposition" in the formulae-as-types approach are in fact constructions which result from proofs (seen as processes); we shall call them *proof-objects*.

Our analysis of the notion of a basic domain etc. is only partial, and leaves us with several awkward questions (even if the analysis suffices for the connections with realizability in section 5). As to the method for resolving these difficulties proposed by Martin-Löf, see the end of section 3.

Combining the formulae-as-types approach with the idea that basic equalities, if provable at all, have fixed canonical proofs (that is to say, a proof of an equality carries no relevant mathematical information besides its truth), we obtain interpretations of logic of the realizability type.

This is demonstrated in particular for (the lowest level of the) systems of ML 75 and ML 82: after a description of Martin-Löf's systems at the lowest level (Section 4), seen as calculi of basic domains, and the theory **APP** of applicative systems (section 5), we describe an interpretation of these systems in **APP** under which the logic corresponds to the logic in **APP** under abstract realizability (section 6). Section 7 is devoted to some supplementary discussion. A good deal of the paper is expository, and we lay little claim at mathematical novelty, except for the fact that the connections with realizability are made more explicit than hitherto. The proofs of Renardel's results (5.4, 6.3) will be published elsewhere.

1.6. *Acknowledgements*. Apart from the written sources (the principal indebtedness being to Beeson 1980, Sundholm 1983, ML 82) we have to thank Robin Grayson for helpful remarks, Gerard Renardel for his willingness to prove the theorems proposed by us, and especially Peter Aczel for conversation and correspondence – which does not mean that he is in any way to be held responsible for the errors of our ways.

2. HEYTING'S EXPLANATIONS OF THE
INTUITIONISTIC LOGICAL OPERATIONS

2.1. For future reference, we recall the well-known clauses of the explanation of the logical constants along the lines of Heyting 1934. The explanations tell us what a proof of a compound statement is in terms of proofs for the constituents. Thus

(H1)   A proof of $A \wedge B$ is given by presenting proofs of A and B.

(H2)   A proof of $A \vee B$ is given by presenting either a proof of A or a proof of B.

(H3)   A proof of $A \to B$ is given by a construction (function, operation) transforming every proof of A into a proof of B.

(H4)   no proof establishes $\perp$(absurdity).

We explain the quantifiers ranging over a "basic domain" D; a basic domain is such that "d belongs to basic domain D" does not need further proof; from d itself it is evident that it belongs to D (section 3 discusses the concept of basic domain at length). Constructivists generally agree that at least $\mathbb{N}$, the set of natural numbers is a basic domain.[4]

(H5)   A proof of $\forall x \in D \, A(x)$ is given by a construction f (function, operation) transforming any $d \in D$ into a proof fd of $A(d)$.

(H6)   A proof of $\exists x \in D \, A(x)$ is given by presenting $d \in D$ and a proof of $A(d)$.

Remarks.

(a) In these explanations, "p is a proof of A", "p proves A" and "p is a construction-process establishing A" are regarded as synonymous.

(b) Heyting does not explicitly distinguish between basic domains and other possible domains of quantification; the restriction is necessary, however. If, for example, D was defined as a subset of $\mathbb{N}$ by means of a predicate essentially involving an existential quantifier, so $x \in D \equiv \exists y \, D'(x, y)$, then the constructions, in proofs of $\forall x \in D \, A(x)$, which is equivalent to $\forall x \forall y (D'(x, y) \to A(x))$, cannot in general depend on x as element of D only, but also on y and proofs of $D'(x, y)$, information extracted from a proof of $x \in D$.

(c) In clauses H3, H5 we must of course also *recognize* that the operation has the required property; this leads to the issue of "second clauses" discussed in the next subsection.

(d) In clauses H3, H4 the function concept enters in the explanations.

2.2. The explanations H1-5 have been taken by Kreisel (1960, 1965) as a starting point for a theory of constructions (in the expectation that this could lead to technically useful formal interpretations). In Kreisel's theories, to the operation in the clauses H3, H5 is joined a construction (proof) establishing that the operation has the property required[5].

Later Kreisel observed (1970, pp. 145–146, in note 11): "There is an additional distinction which has so far not been formally necessary, but which is probably important, for example in the explanation of implication (or universal quantification). When we think of the pair (p_1, p_2),

> p_1 proves the identity: for variable p, if p proves A then $p_2(p)$, i.e., p_2 applied proves B,

p_2 is a genuine function or operation, while p_1 recognizes that p_2 satisfies the condition stated; thus p_1 is a judgment. But similarly, since in general both the arguments p and the values $p_2(p)$ of p_2 are such pairs, say $p = (p', p'')$ and $p_2(p) = (p_2', p_2'')$, should the function p_2 act both on p'' and p' (or only on p'')? It is perhaps significant that in Gödel's interpretation (1958), of $\mathfrak{A} \to \exists s \mathfrak{B}$ for purely universal $\mathfrak{A}$ and $\mathfrak{B}$, we have $\exists s(\mathfrak{A} \to \mathfrak{B})$, that is, the function does not depend on the (hypothetical) proof of $\mathfrak{A}''$.

In Kreisel's theories of 1960 and 1965, the judgment-part of a proof is regarded as a *mathematical* object. Sundholm (1983) analyzes the matter in detail and argues that this is mistaken: the judgment itself should not be treated as a mathematical object. In the next subsection we summarize some points of Sundholm's paper which are relevant to what follows.

2.3. *Summary of Sundholm* (1983). Sundholm lists a number of basic assumptions underlying the discussions of the theory of proof and constructions, and the interpretation of intuitionistic logic in the literature; we list a few that are directly relevant to our discussion here:

(I)   Mathematical propositions are seen to be true (may be asserted) by being proved;

(II)   Proofs must be understood as such in order to be proofs;

(III)   The assertion of a proposition is not itself a proposition;
(IV)   Proofs are constructions;
(V)   Constructions are mathematical objects.

(I) and (II) may be regarded as noncontroversial among constructivists. Heyting held to (III) and Sundholm agrees. Both Heyting and Kreisel, in their writings cited, adhere to (IV) and (V) (on Sundholm's reading). Sundholm notes that there are difficulties in reconciling (I–V), and makes a proposal to get out of the tangle: he distinguishes between constructions-as-processes and constructions that are the result of a process, i.e., two different meanings of the word "construction" are distinguished; in the foregoing, we have at several places already made this distinction explicit by our choice of wording. Only constructions in the second sense are to be regarded as mathematical objects. A *proof* is a construction *process*, yielding a construction as a result; mentally following the process convinces us of the proof (i.e., makes us accept the truth of the proposition proved).

The *result* of a proof (construction-process) one might call a *proof-object* ("witness to the truth"). The proof is more than just a proof-object.

So far, we have summarized Sundholm. All this fits in well with the quotation from Kreisel above; the judgment is by Sundholm regarded as not being a mathematical object, and then it seems quite logical to let the mathematical components (proof-objects) of proofs of implications and universal statements, i.e., the operations mentioned in H3, H5, act only on proof-objects as arguments.

2.4. To what extent does a proof-object determine a proof? In other words, presented with a proof-object, can we construct a proof, that is, can we mentally follow a (proof-) process that results in the given proof object?

In simple situations this is certainly possible; from the proof-objects $\lambda ab.a$, $\lambda abc.ac(bc)$ corresponding to proofs of $A \to (B \to A)$, $(A \to (B \to C)) \to ((A \to B) \to (A \to C))$ we can easily recover complete proofs. In general, this is not plausible any more. Consider for example the following situation: if t, t' are closed numerical terms, $t = t'$ is *formally* proved by evaluating t, t' (i.e., showing that they can be equated to numerals $\bar{n}$, $\bar{m}$ say), and then comparing $\bar{n}$, $\bar{m}$. However, a proof of $t = t'$ as a mental construction (-process) with a corresponding proof-object concerns the *objects* n, m denoted by the terms t, t'. Comparison of these objects is immediate and carries no mathematical information beyond the truth of the equation; so we may as well denote such a trivial canonical proof by an arbitrary fixed object, say 0. A proof of $\forall x(t[x] = 0)$, t a numerical term with parameter x, convinces us of the fact that $\lambda x.0$ is a function which assigns to each $x \in \mathbb{N}$ a proof-object of $t[x] = 0$, hence $\lambda x.0$ is a proof-object for $\forall x(r[x] = 0)$, from which we cannot reconstruct the possibly quite complicated argument showing that $\forall x(t[x] = 0)$.

In this connection it is to be noted that

(A) The feature just signalled is well-known from realizability interpretations.

(B) In intuitionistic mathematical practice proofs of propositions under a premiss in Π_1^0-form do not actually depend on the proof of the premiss, but only on its truth.

This is reminiscent of the weak independence-of-premiss principle valid under Gödel's Dialectica interpretation:

IP$_0$ $[A \to \exists yBy] \to \exists y[A \to By]$ (y not free in A)

with A in Π_1^0-form. There is a difference, however: IP$_0$ requires the construction of y as soon as we know $A \to \exists yBy$; the principle above only requires the construction of a y *provided* A is true, the y not depending on the proof of A. Combined with choice principles we can make this visible formally: IP$_0$ in strengthened form yields

$$\forall x[Ax \to \exists y \, B(x, y)] \to \exists f : \mathbb{N} \to \mathbb{N}\forall x[Ax \to B(x, fx)]$$

where the independence from the proof of Ax yields

$$\forall x \in A \exists yB(x, y) \to \exists f : A \to \mathbb{N}\forall x \in AB(x, fx)$$

(f is not required to be total on $\mathbb{N}$).

(C) Compare also the fact that for sentences without $\to$ in arithmetic, truth automatically leads to the insight that a canonical infinitary proof-tree exists, in a system with the ω-rule instead of induction; the infinitary proof-tree itself becomes a proof only with additional information, establishing that the proof-tree is correct as such. In particular, a proof must provide a method by which the infinitely many top nodes are seen to be true, and the proof-tree is seen to be well-founded.

The divergence between proof-objects and the informal arguments corresponding to formal proofs of the usual kind is at first sight unsatisfactory: after all, verifying that 2^7-1 and "the least prime following 13 consecutive composite numbers" both refer to 127 requires some work, we have to *prove* something. Cf. also section 3.3.

3. BASIC DOMAINS AND IDENTITY

3.1. *The idea of a basic domain.* We set out with the following explanations concerning the relations between the notions of "basic domain" "(basic) equality on a basic domain", "mathematically relevant" and "proof-object" (these *explanations* by no means define the notions concerned):

(1) A *basic domain* is a domain such that there is a "*basic equality*" between elements of the domain; two objects of such a domain are regarded as equal (identical) if and only if they agree in all aspects relevant in any potential mathematical context ("agree in all (potentially) relevant mathematical aspects").

(2) Objects d belonging to a basic domain D may themselves be regarded as canonical proof-objects for the proof that d belongs to D.

We paraphrase and spell out some consequences of (1) and (2) in the comment below.

Comments.

(a) Suppose D to be a domain, $d \in D$, and suppose that $d \in D$ can be established in two essentially different ways resulting in distinct proof-objects p, p'; proof-object p of a proof establishing $d \in D$ is a mathematically relevant aspect of d, and should be given together with d if D is basic; in general, there corresponds to D a basic domain D' of pairs (d, p), with p a proof-object for $d \in D$.

(b) As a consequence of (1), we expect that all operations Φ a basic domain D respect its basic equality $=_D$; for if not, $\Phi a \neq \Phi b$ would indicate a mathematically relevant difference for a, b.

(c) Condition (2) is connected with our wish to regard propositions as given by the collection of proof-objects for the proposition; operations on proofs of propositions are actually operations on proof-objects. Propositions thus become special basic domains, or one might say that basic domains are generalized propositions. ($\mathbb{N}$ however is an example of a basic domain not normally considered to represent a proposition, though one might consider $n \in \mathbb{N}$ as a proof-object for "$\mathbb{N}$ is inhabited").

For basic domains A, B, the proof-objects of the implication $A \to B$

are mappings from A to B; and if for each $x \in A \, B[x]$ is a basic domain, the proof-objects of $\forall x \in A.B[x]$ are functions f with domain A, such that $f(x) \in B[x]$ for $x \in A$. As observed before, the notion of proof-object of an implication or universal statement depends on the notion of function.

(d) Thus the explanations still leave freedom to impose extra requirements on basic equalities; but restrictions on the notion of basic equality entail corresponding restrictions on the notion of operation from basic domain to basic domain, and hence also of what shall be regarded as "mathematically relevant" and on what shall be taken to be a legitimate proof-object of an implication of a universal statement.

(e) The following axiom of choice holds for basic domains D, D':

$$\forall x \in D \exists y \in D' A(x, y) \to \exists F : D \to D' \forall x \in D A(x, Fx);$$

for a proof-object of the premiss is a mapping f assigning to $d \in D$ a proof-object fd of $\exists y \in D' A(d, y)$. As fd is given by $d' \in D'$ and a proof of $A(d, d')$, we can extract the required F from f.

3.2. *An example.* If D is a basic domain, and we choose two times an element (not: two elements) of D, say d_0, d_1, then $D' \equiv \{d_0, d_1\}$ (with equality inherited from D) cannot be regarded as a basic domain, unless $d_0 = d_1 \vee d_0 \neq d_1$; let $D'' \equiv \{\langle d_0, 0 \rangle, \langle d_1, 1 \rangle\}$, then $\forall x \in D' \exists y \in \{0, 1\}$ ($\langle x, y \rangle \in D''$), but choice would give f such that $\forall x \in D'(\langle x, fx \rangle \in D'')$, and thus $d_0 = d_1 \leftrightarrow fd_0 = fd_1$.

Choosing d_0, d_1 from D has added relevant information (the choice-number) in connection with their membership of D'. Now it is rather D'' that plays the rôle of a basic domain.

The argument *does* show that basic domains indexed by natural numbers ought to have a decidable equality.

Should basic equality always be decidable? After section 5 it will have become clear that we can take realizability predicates for numerical realizability with equality from $\mathbb{N}$, as a model of the theory of basic domains in which basic equalities are decidable, and realizability predicates for functional realizability, with extensional equality between functions, as a model where this is not the case.

3.3 *The theory of types.* The discussion in section 2 and subsections 3.1, 3.2 above are an attempt to clarify the relationships between the informal (intuitive) notions of basic domain, basic equality, proof and proof-object, by approaching them from several different angles. This led us to the insight that these notions hang together, and that certain relations between them were forced upon us. On the other hand, we are left with several awkward questions. First of all, we have in no way characterized "basic domain", "basic equality" – how can we introduce a basic domain?

Secondly, on the formulae-as-types interpretation of logic, what about the divergence between formal proofs and proof-objects we signalled at the end of section 2?

Type theory in the sense of Martin-Löf can be seen as an attempt to answer these questions. We shall give a brief sketch here, primarily based on ML 82 and lectures given by Martin-Löf at Münich, autumn 1980, and to a lesser extent on Dummett 1977, Sundholm 1982; for the rest we must refer the reader to these sources.

A *category* is explained by saying how to obtain elements of the category, and what it means for two elements of a category to be equal. (Category is not used in the sense of category-theory, but derives from philosophical terminology). Examples of categories: types, elements of a type, families of types over a type, etc.

A *type* is explained by telling what its (canonical) elements are and how equal (canonical) elements are formed. Perhaps due to Martin-Löf's conviction that objects cannot be handled without their linguistic descriptions, "canonical expression" and "canonical object" seem to be used virtually interchangeably in ML 82. Typically, the form of canonical elements of a type is specified by *introduction* rules.

Thus the type of natural numbers $\mathbb{N}$ is explained by saying that its elements are 0 or successors of elements of $\mathbb{N}$.

$$0 \in \mathbb{N}, \qquad \frac{t \in \mathbb{N}}{\mathbf{suc}(t) \in \mathbb{N}}$$

These rules reflect the way natural numbers are generated.

Similarly, given two types A, B, the conjunction or product type $A \wedge B$ (or $A \times B$) has pairs as elements, corresponding to an introduction rule

$$(\wedge \mathrm{I}) \qquad \frac{t \in A \quad t' \in B}{\langle t, t' \rangle \in A \wedge B}$$

As a counterpart to introduction rules there are elimination rules which *exploit* the fact that the elements of a type are generated by the introduction rules. Thus the elimination rule for $\mathbb{N}$ is nothing but induction and definition by recursion (combined into a single rule); and for $A \wedge B$ we have

$$(\wedge E) \qquad \frac{t \in A \wedge B}{\mathbf{j}_1 t \in A} \quad \frac{t \in A \wedge B}{\mathbf{j}_2 t \in B};$$

$\mathbf{j}_1 t, \mathbf{j}_2 t$ are noncanonical expressions. We can stipulate rules for giving a canonical expression as value to noncanonical expressions. If it has a canonical value $\langle t', t'' \rangle$, the value of $\mathbf{j}_1 t$ is equal to the value of t' etc.

Compare this to ordinary arithmetic: the canonical presentation of numbers is by numerals; but definitions, in particular by recursion, permit us to regard other closed terms such as $2^7 + 3$ as descriptions of numbers to which we can assign a numeral as value.

Canonical expressions for proofs of implications must have the form of functions, and implication introduction is the rule for generating (canonical expressions for) proof-objects of implications. "Modus ponens" is its elimination counterpart justified by the insight that an implication proof should be a function.

Dummett proposed a distinction between "canonical proofs" (represented by canonical expressions for proof-objects in Martin-Löf's theory) and arbitrary (indirect) proofs, as a way of exluding a trivial reading of Heyting's explanations.

For suppose the proof-objects were taken to be proofs in some formal system containing modus ponens, and let p prove $A \to B$, q prove A, while $\varphi(p, q)$ is the proof of B resulting from p and q by modus ponens; whatever the operation φ and the proof p are, $\lambda x.\varphi(p, x)$ represents a proof-function which to any proof q of A assigns a proof of B and is therefore a proof of $A \to B$ [which is already implicit in p]. (This same idea can be used for a trivial modelling of Kreisel's theory of constructions, cf. 7.3.) The distinction between direct and indirect proofs removes this trivial (and in a sense, circular) interpretation: Heyting's explanations must be taken to concern direct proofs. This agrees with the distinction between canonical and non-canonical expressions in Martin-Löf's theories; the divergence between proof-objects and formal proofs is connected with the distinction between direct and indirect proofs in Dummett's sense.

4. MARTIN-LÖF'S SYSTEMS

4.1. We shall introduce in this section two theories $\mathbf{ML}_0$ and $\mathbf{ML}_0^i$ corresponding to the lowest level of Martin-Löf's formalisms. The axioms are easily motivated by thinking of them (naively) as describing a theory of basic domains; the following principles for generating basic domains are assumed:

(B1) $\mathbb{N}$ is a *primitive* domain; primitive domains are basic domains.

Let $\{B_x : x \in B\}$ be *a family of basic domains B_x indexed by elements x of a basic domain A* (the italicized expression is regarded as a primitive notion), then

(B2) $\Pi x \in A.B_x$ is the basic domain consisting of all functions f with domain A and $fx \in B_x$, for all $x \in A$;

(B3) $\Sigma x \in A.B_x$ is the basic domain consisting of all pairs $\langle x, y \rangle$ with $x \in A$, $y \in B_x$.

Finally we introduce

(B4)   For any basic domain A, and elements $a \in B$, $b \in A$ there is a basic domain $I(A, a, b)$ which contains a single canonical element r iff $a = b$ in A. (I.e. in conventional notation $I(A, a, b) = \{r : a = b\}$.)

The theories of ML 75, ML 82 reckon with further basic domains; of these the k-element domains $\mathbb{N}_k$ ($k \in \mathbb{N}$) and the disjoint-sum formation $A + B$ (from basic domains, A, B) are in fact explicitly definable (i.e., we can construct appropriate basic domains from $\mathbb{N}$, Π, Σ, I which behave like $\mathbb{N}_k$, $A + B$) and thus we shall leave them out here, so as to keep the formal apparatus simple. Other primitive domains and operations for constructing new basic domains from given ones strengthen the system, and will not be considered in this paper.

4.2. *Logic in terms of basic domains: formulae-as-types.* Each basic domain (or *type*) A can be seen in a double rôle:

(i) A is a *set*, with *elements* $a \in A$; but also

(ii) A is a (generalized) *proposition*, with proof-objects $a \in A$.

The two rôles are connected: an element b of the set A may be seen as being itself a canonical proof-object for a proof that b belongs to A (A holds).

Logic in the usual sense can be expressed in terms of basic domains as follows (a recapitulation of the formulae-as-types idea): for B not depending on x, $\Pi x \in A.B$ and $\Sigma x \in A.B$ represent the proof-objects of $A \to B$, $A \land B$ respectively. If B does depend on x, $\Pi x \in A.B$ and $\Sigma x \in A.B$ represent the proof-objects of $\forall x \in A.B(x)$, $\exists x \in A.B(x)$.

Proof-objects for equalities between elements of a basic domain are supposed not to contain relevant mathematical information beyond the truth of the equality. Thus all proof-objects of such an equality may be supposed to be represented by a single canonical proof-object $\mathbf{e}$.

In the theory $\mathbf{ML}_0$ corresponding to the basic theory of ML 82, equality at each type is interpreted as extensional equality; in the "intensional" theory $\mathbf{ML}_0^i$ extensionality is not imposed; there it is easier to think of equality as a primitive satisfying certain minimum requirements. $\mathbf{ML}_0^i$ is close to, but not the same as the lowest level theory of ML 75.

4.3. *Description of the system* $\mathbf{ML}_0^i$. We shall not be overly formal in our description[6] of the system; more detailed presentations are easily constructed by looking at ML 82 or Diller 1980.

All statements in the theory are of the forms "A type", "$t \in A$", "$t = s \in A$", "$A = B$". Such statements are derived under (possibly empty) finite sets of assumptions called contexts; contexts can be arranged as

(*) $x_1 \in A_1, x_2 \in A_2[x_1], x_3 \in A_3[x_1, x_2], \ldots x_n \in A_n[x_1 \ldots x_{n-1}]$

(all variables shown). Let us use $\Gamma, \Gamma'. \ldots$ for contexts. A new assumption $x_{n+1} \in A_{n+1}[x_1, \ldots, x_n]$ may be added only to a context Γ as in (*), when we first have derived

$$\Gamma \Rightarrow A_{n+1}[x_1, \ldots, x_n] \text{ type.}$$

(which means that $A_{n+1}[x_1 \ldots x_n]$ is a type ($\approx$ basic domain) under the assumptions of Γ).

Intuitively, $t = s \in A$ may be read as: t, s are equal elements of type

A. A more suggestive notation would have been $t =_A s$. We have the following rules (1)–(11):

(1) $\Gamma \Rightarrow \Theta$ for $\Theta \in \Gamma$

(2) $$\frac{\Gamma \Rightarrow \Theta}{\Gamma \cup \Gamma' \Rightarrow \Theta} \text{ (thinning)}$$

(3) obvious substitution rules such as

$$\frac{\Gamma' \Rightarrow t \in A \quad \Gamma, x \in A \Rightarrow B[x] \text{ type}}{\Gamma \cup \Gamma' \Rightarrow B[t] \text{ type}} \text{ etc.}$$

For the rest of the rules we show contexts and parameters where necessary only.

(4) $\mathbb{N}$ type; and for x not in Γ

$$\frac{\Gamma, x \in A \Rightarrow B[x] \text{ type}}{\Gamma \Rightarrow \Pi x \in A . B[x] \text{ type}}, \quad \frac{\Gamma, x \in A \Rightarrow B[x] \text{ type}}{\Gamma \Rightarrow \Sigma x \in A . B[x] \text{ type}} \text{ and}$$

$$\frac{A \text{ type} \quad t \in A \quad s \in A}{I(A, t, s) \text{ type}}.$$

Abbreviations: if x is not free in B, we write

$$A \rightarrow B \equiv \Pi x \in A . B, \quad A \wedge B \equiv \Sigma x \in A . B.$$

(5) we have combinators at all suitable types

$$\frac{A \text{ type} \quad B \text{ type}}{\mathbf{k} \in A \rightarrow (B \rightarrow A)},$$

$$\frac{A \text{ type} \quad B \text{ type} \quad C \text{ type}}{\mathbf{s} \in (A \rightarrow (B \rightarrow C)) \rightarrow ((A \rightarrow B) \rightarrow (A \rightarrow C))}.$$

(We may in fact assume these combinators to be notationally distinguished by appending the relevant types: $\mathbf{k}^{A,B}$, $\mathbf{s}^{A,B,C}$.) The combinators permit us in the usual way to define λ-abstraction.

(6) Π-rules

(III) $$\frac{\Gamma, x \in A \Rightarrow b \in B}{\Gamma \Rightarrow \lambda x . b \in \Pi x \in A . B} \quad (x \text{ not free in } \Gamma);$$

(ΠE) $$\frac{t \in \Pi x \in A.B[x] \quad t' \in A}{tt' \in B[x]}.$$

(7)   Σ-rules

(ΣI) $$\frac{t \in A \quad t' \in B[t]}{\langle t, t' \rangle \in \Sigma x \in A.B[x]}$$ (for $\langle t, t' \rangle$ we can also write $\mathbf{j}tt'$);

(ΣE) $$\frac{t \in \Sigma x \in A.B[x]}{\mathbf{j}_1 t \in A} \,; \qquad \frac{t \in \Sigma x \in A.B[x]}{\mathbf{j}_2 t \in B[\mathbf{j}_1 t]}$$

(8) I-rules

(II) $$\frac{a = b \in A}{\mathbf{e} \in I(A, a, b)}\,; \qquad \frac{c \in I(A, a, b)}{c = \mathbf{e} \in I(A, a, b)}\,;$$

(IE) $$\frac{c \in I(A, a, b)}{a = b \in A}$$

(9)   N-rules

(NI) $0 \in \mathbb{N}$; $$\frac{t \in \mathbb{N}}{\mathbf{suc}(t) \in \mathbb{N}}\,;$$

(NE) $$\frac{t \in \mathbb{N} \quad t' \in A[0] \quad t'' \in \Pi x \in \mathbb{N}(A[x] \to A[\mathbf{suc}(x)])}{\mathbf{r}t't''t \in A[t]}.$$

Instead of the premiss "$t'' \in \ldots$", we might also have used the more natural $x \in \mathbb{N}, y \in A[x] \Rightarrow t^*(x, y) \in A[\mathbf{suc}(x)]$ with conclusion $\mathbf{r}t'(\lambda xy.t^*)t \in A[t]$.

(10) $=$ between elements of a type, and between types is reflexive, symmetric, transitive, permits replacement, e.g.,

$$\frac{a \in A \quad A = B}{a \in B}\,, \qquad \frac{b = c \in B \quad A[b]\,\text{type}}{A[b] = A[c]}$$

(11)   All the usual combinator axioms hold at all appropriate types ($t = s$ short for: $t = s \in A$ for appropriate A), i.e.,

$$\mathbf{k}xy = x, \qquad\qquad \mathbf{s}xyz = xz(yz)$$
$$\mathbf{r}xy0 = x, \qquad\qquad \mathbf{r}xy(\mathbf{suc}(z)) = yz(\mathbf{r}xyz)$$
$$\mathbf{j}_1\langle x, y \rangle = x, \qquad\quad \mathbf{j}_2\langle x, y \rangle = y.$$

4.4. *Comparison between* $\mathbf{ML}_0^i$ *and* $\mathbf{ML}_0^*$. $\mathbf{ML}_0^*$ is the system of ML 75 at lowest level, i.e., with all references to the universes V_n deleted; this is the "strong system" of Diller 1980 (the λ-abstraction there might also be understood in a defined sense as in ML 75).

$\mathbf{ML}_0^i$ is similar, but there are several differences.

(A) ML 75 uses schemata to introduce function constants; we use combinatorial constants $\mathbf{k}, \mathbf{s}, \mathbf{r}$ (and implicitly a constant for application) instead. This difference is not a very essential one, but rather one of technical convenience. For given types A, B, C the combinators $\mathbf{k}, \mathbf{s}$ may be interpreted in $\mathbf{ML}_0^*$ by certain function constants introduced as proofs of $A \rightarrow (B \rightarrow A)$, $(A \rightarrow (B \rightarrow C)) \rightarrow ((A \rightarrow B) \rightarrow (A \rightarrow C))$ respectively. Similarly, the recursion *schema* at all types is available in $\mathbf{ML}_0^*$. At any given type, a term playing the rôle of the recursor $\mathbf{r}$ can be obtained in $\mathbf{ML}_0^*$ in a routine fashion using the recursion schema at a suitable more complex type.

(B) A much more essential difference is the following. ML 75 and ML 82 both use identity types; $p \in I(A, t, s)$ is intuitively read as "p establishes $t = s$ at type A". However, in $\mathbf{ML}_0^*$ we do not have expressions $t = s \in A$, but only "t conv s"; here "conv" is a conversion relation which syntactically generates a notion of equality by definition: t conv s implies that t, s stand for (denote) the same object. However, we do not have the means to say that $t = s \in A$ if and only if t and s are equal by definition. On the other hand, in $\mathbf{ML}_0^i$ we also have expressions $t = s \in A$; one might think of these as generated by adding to $\mathbf{ML}_0^*$ a rule

$$(1) \qquad \frac{p \in I(A, t, s)}{t = s \in A}$$

In $\mathbf{ML}_0^*$, if t conv s, then also $I(A, t, t)$ conv $I(A, t, s)$, and thus, since $I(A, t, t)$ has a canonical proof $r(t)$ (i.e., $r(t) \in I(A, t, t)$), combination with the following rule of $\mathbf{ML}_0^*$

$$\frac{a \in A \quad A \text{ conv } B}{a \in A}$$

yields a derived rule

$$\frac{t \text{ conv } s}{r(t) \in I(A, t, s)}.$$

If we have a term model of closed normal terms[7] in mind, then $I(A, t, s)$

for closed normal terms t, s can be interpreted as literal identity, and the only possible *proof-object* of $I(A, t, s)$ is always of the form $r(t)$. Dropping the dependence on t (an inessential feature) and writing $\mathbf{e}$ for $r(t)$, this leads to inclusion not only of $\mathbf{e} \in I(A, t, t)$ but also a converse

$$(2) \qquad \frac{t' \in I(A, t, t)}{t' = \mathbf{e} \in I(A, t, t)}$$

which says that there is only one canonical proof of an identity. In $\mathbf{ML}_0^*$, (1), (2) would correspond to

$$\frac{p \in I(A, t, s)}{t \text{ conv } s} \qquad \frac{p \in I(A, t, t)}{p \text{ conv } \mathbf{e}}$$

which is in conflict with starting point of ML 75 that "conv" should be syntactically determined.

In a similar[8] way rules such as $\dfrac{c \in N_1}{c \text{ conv } 1}$, expressing that N_1 has only a single element, are missing from $\mathbf{ML}_0^*$.

Thus the transition from $\mathbf{ML}_0^*$ to $\mathbf{ML}_0^i$ involves a double move: making explicit the realizability character (equalities are established by a unique object) and replacing the purely syntactic "conv" by a primitive equality $=$ at each type.

Technically, $\mathbf{ML}_0^i$ is a convenient intermediate system between $\mathbf{ML}_0^*$ and $\mathbf{ML}_0$, permitting us to demonstrate the realizability aspect of type theory in the simplest situation.

4.5. *The extensional theory* $\mathbf{ML}_0$. The description of this theory is now relatively easy: most of the rules correspond to those of $\mathbf{ML}_0^i$, except that now we primarily give rules for "$t = s \in A$" generating an equivalence relation at each type; "$t \in A$" may in fact be regarded as short for "$t = t \in A$". Rules (1)–(5) as before[9];

$$(6) \qquad \frac{\Gamma, x = x \in A \Rightarrow t = t' \in B}{\Gamma \Rightarrow \lambda x.t = \lambda x.t' \in \Pi x \in A.B} \qquad (x \text{ not free in } \Gamma);$$

$$\frac{\Gamma \Rightarrow t \in \Pi x \in A.B}{\Gamma \Rightarrow \lambda x.tx = t \in \Pi x \in A.B} \qquad (x \text{ not free in } t);$$

$$\frac{t = t' \in A \quad t_1 = t_1' \in \Pi x \in A.B[x]}{t_1 t = t_1' t' \in B[t]}$$

(7)
$$\frac{t = t' \in A \quad t_1 = t_1' \in B[t_1]}{\langle t, t_1 \rangle = \langle t', t_1' \rangle \in \Sigma x \in A.B[x]},$$

$$\frac{t = t' \in \Sigma x \in A.B[x]}{\mathbf{j}_1 t = \mathbf{j}_1 t' \in A}, \qquad \frac{t = t' \in \Sigma x \in A.B[x]}{\mathbf{j}_2 t = \mathbf{j}_2 t' \in B[\mathbf{j}_1 t]}.$$

(8)   as before.

(9) $0 = 0 \in \mathbb{N}$, $\dfrac{t = t' \in \mathbb{N}}{\mathbf{suc}(t) = \mathbf{suc}(t') \in \mathbb{N}}$;

$$\frac{t = t' \in \mathbb{N} \quad t_1 = t_1' \in A[0] \quad t_2 = t_2' \in \Pi x \in \mathbb{N}(A[x] \rightarrow A[\mathbf{suc}(x)])}{\mathbf{r} t_1 t_2 t = \mathbf{r} t_1' t_2' t' \in A[t]}.$$

(10), (11) similarly to before; now equality between types is interpreted extensionally throughout.

5. APPLICATIVE SYSTEMS

5.1. *The theory* **APP**. We first briefly describe the theory **APP** of applicative systems along the lines of Feferman 1979, with a slight difference: where Feferman uses a ternary predicate App(x, y, z) expressing: "x is applicable to y with result z" we shall regard application (simply denoted by juxtaposition) as a partially defined binary function; thus from the constants and the variables of the theory we can form terms which are not always defined. Accordingly, we have to adopt logic with an existence predicate E (as in Scott 1979; only we shall always assume our variables to exist). This necessitates two changes in the usual axiomatization of intuitionistic logic: the quantifier axioms $\forall x Ax \rightarrow At$ and $At \rightarrow \exists x Ax$ have now to be replaced by

$$\forall x Ax \wedge Et \rightarrow At \qquad At \wedge Et \rightarrow \exists x Ax,$$

or in a natural-deduction form

$$(\forall E) \quad \frac{\forall x Ax \quad Et}{At} \qquad (\exists I) \quad \frac{At \quad Et}{\exists x Ax}.$$

APP is a single-sorted theory with variables $(x, y, z, u, v, w \ldots)$, individual constants 0, **suc**, **k**, **s**, $\mathbf{j}_1$, $\mathbf{j}_2$, **j**, **d**, **prd**, a two place function constant **app**, predicate constants $=$, E, $\mathbb{N}$.

Terms: all individual constants and variables are terms; if t_1, t_2 are terms, so is **app** (t_1, t_2), abbreviated to $t_1 t_2$. We use the convention of associating to the left: $(\ldots ((t_1 t_2) t_3) t_4 \ldots t_n)$ is abbreviated as $t_1 t_2 t_3 \ldots t_n$.

Prime formulae: if t_1, t_2 are terms, then $t_1 = t_2$, $E t_1$ and $\mathsf{N}(t_1)$ (also written $t_1 \in \mathsf{N}$) are prime formulae.

Formulae are obtained from prime formulae with the help of logical operations.

Axioms

(I) $\quad \begin{cases} E x \text{ for any variable } x, \\ E t \leftrightarrow t = t, \\ t = t' \to t' = t, \ t = t' \wedge t' = t'' \to t = t''. \end{cases}$

Defining equivalence $\simeq$ by

$$t \simeq t' \equiv (E t \vee E t') \to t = t';$$

we further require

(II) $\quad \begin{cases} t \simeq t' \to t'' t \simeq t'' t' \wedge t t'' \simeq t' t'' \\ t \in \mathsf{N} \to E t, \ t = t' \wedge E t \to E t', \ t \in \mathsf{N} \wedge t = t' \to t' \in \mathsf{N} \\ E t t' \to E t \wedge E t'. \end{cases}$

Combinatorial axioms

(III) $\quad \begin{cases} \mathbf{k} t y \simeq t, \quad \mathbf{E s} x y, \quad \mathbf{s} t_1 t_2 t_3 \simeq t_1 t_3 (t_2 t_3) \quad \text{(combinators)} \\ \mathbf{E j} x y, \quad \mathbf{j}_1 (\mathbf{j} t y) \simeq t, \quad \mathbf{j}_2 (\mathbf{j} x t) \simeq t \quad \text{(pairing)} \\ x, y \in \mathsf{N} \to \mathbf{d} x y t t' \simeq \quad t \text{ if } x = y \wedge E t' \quad \text{(definition} \\ \qquad\qquad\qquad\qquad\qquad\quad t' \text{ if } x \neq y \wedge E t \quad \text{ by cases)} \end{cases}$

Natural numbers

(IV) $\quad \begin{cases} 0 \in \mathsf{N}, \quad t \in \mathsf{N} \to \mathbf{suc}(t) \in \mathsf{N} \wedge \neg \mathbf{suc}(t) = 0, \\ \mathbf{prd}(0) = 0, \quad \forall x \in \mathsf{N}(\mathbf{prd}(x) \in \mathsf{N} \wedge \mathbf{prd}(\mathbf{suc}(x)) = x) \\ \qquad\qquad\qquad\qquad\qquad\qquad\qquad\qquad\qquad \text{(predecessor)}, \\ \text{induction axiom.} \end{cases}$

From the combinators we define λ-abstraction in a standard way such that $\mathbf{APP} \vdash E \lambda x.t$ for every term t; because of the possibility of self-application we can define a term $\mathbf{r}$, the recursor, such that

$$\mathbf{r} x y 0 \simeq x$$
$$\mathbf{r} x y \mathbf{suc}(z) \simeq y z (\mathbf{r} x y z)$$

and thus we can construct, e.g., all primitive recursive functions and functionals in **APP**.

5.2. *Embedding* **HA**, **N-HA**$^\omega$ *in* **APP**.**HA** can be embedded in an obvious way in **APP**: the recursor **r** gives us the required primitive recursive functions on $\mathbb{N}$, and numerical quantifiers correspond to restricted quantifiers $\forall x \in \mathbb{N}$, $\exists x \in \mathbb{N}$ in **APP**.

This is easily extended to an embedding $^+$ in **APP** of **N-HA**$^\omega$, intuitionistic finite type arithmetic with a primitive notion of equality at all finite types (cf. Troelstra 1973, 1.6). We write $\mathbb{N}_0 \equiv \mathbb{N}$, and if we have already defined sets $\mathbb{N}_\sigma$, $\mathbb{N}_\tau$ in **APP** as our interpretation of the range of quantifiers of type σ, τ, the ranges for types $(\sigma)\tau$ (or $\sigma \to \tau$) and $\sigma \times \tau$ are defined as

$$\mathbb{N}_{(\sigma)\tau} = \{x : \forall y \in \mathbb{N}_\sigma(xy \in \mathbb{N}_\tau)\},$$
$$\mathbb{N}_{\sigma \times \tau} = \{\langle x, y \rangle : x \in \mathbb{N}_\sigma \wedge y \in \mathbb{N}_\tau\}$$

(which is $\exists$-free and short for $\{z : \mathbf{j}_1 z \in \mathbb{N}_\sigma \wedge \mathbf{j}_2 z \in \mathbb{N}_2 \wedge z = \langle \mathbf{j}_1 z, \mathbf{j}_2 z \rangle\}$). Equations $a =_\tau b$ are mapped by $^+$ into $a^+ = b^+ \wedge a^+ \in \mathbb{N}_\tau$, where a^+ is essentially the term a with all references to types being cancelled. (N.B. $\Pi, \Sigma, R, D', D'', D$ of Troelstra 1973 correspond to $\mathbf{k}, \mathbf{s}, \mathbf{r}, \mathbf{j}_1, \mathbf{j}_2, \mathbf{j}$ here). Then all the axioms and rules of **N-HA**$^\omega$ become valid under this interpretation.

One of the principal models of **APP** is obtained by letting the variables range over $\mathbb{N}$, thought of as the collection of codes of partial recursive functions, and interpreting **app** as partial recursive function application. For $\mathbf{k}$, $\mathbf{s}$, $\mathbf{j}$, $\mathbf{j}_1$, $\mathbf{j}_2$, $\mathbf{d}$ we can take appropriate codes, $\mathbb{N}$ is interpreted as the universe, 0 by itself, and S by (a code of) $\lambda x.x+1$, **prd** by a code of $\lambda x.x \doteq 1$. This leaves arithmetical formulae (essentially) unchanged, and we see that **APP** is conservative over intuitionistic first order arithmetic **HA**.

5.3. *Realizability*. Feferman (1979) defines an abstract version of Kleene's realizability by numbers for **APP**, as follows:

(R1) $tr P \equiv t = 0 \wedge P$ for atomic P,

(R2) $tr A \wedge B \equiv (\mathbf{j}_1 tr A \wedge \mathbf{j}_2 tr B)$,

(R3) $tr A \vee B \equiv (\mathbf{j}_1 t \in \mathbb{N} \wedge ((\mathbf{j}_1 t = 0 \to \mathbf{j}_2 tr A) \wedge (\mathbf{j}_1 t \neq 0 \to \mathbf{j}_2 tr B)))$,

(R4) $tr A \to B \equiv \forall z(zr A \to tzr B) \wedge Et$,

(R5) $tr \forall x Ax \equiv \forall x(txr Ax)$,

(R6) $tr \exists x Ax \equiv \mathbf{j}_2 tr A(\mathbf{j}_1 t) \wedge E(\mathbf{j}_1 t)$.

By induction on the formula-complexity we easily verify the

5.3.1. LEMMA **APP** $\vdash tr A \to Et$.

5.3.2. DEFINITION. A formula A of **APP** or **N-HA**$^\omega$ is $\exists$-free if $\exists$, $\vee$ do not occur in A; A belongs to the class Γ_0 iff in all subformulae $B \to C$ of A, B is $\exists$-free.

The *extended axiom of choice* EAC is the schema

EAC $\forall x[Ax \to \exists y B(x, y)] \to \exists f \forall x[Ax \to Efx \wedge B(x, fx)]$

(A $\exists$-free).

5.3.3. THEOREM (along the lines of Troelstra 1973, 3.6.5, 3.6.6). For all formulae A, $xr A$ is $\exists$-free, and

(i)   **APP** $+$ EAC $\vdash A \Leftrightarrow$ **APP** $\vdash \exists x(xr A)$
(ii)   **APP** $+$ EAC $\vdash A \leftrightarrow \exists x(xr A)$.

For $\exists$-free A, we have (ii) independently of EAC

(iii)   For $\exists$-free A, there is a fixed term t_A such that **APP** $\vdash$ $A \to t_A r A$

(iv)   For $A \in \Gamma_0$, **APP** $\vdash xr A \to A$.

N.B. Feferman 1979 has for R1: $tr P \equiv Et \wedge P$. The change to E-logic (instead of treating E as defined, as Feferman does) is motivated by the fact that this facilitates a rigorous metamathematical treatment of **APP**.

5.3.4. COROLLARY. **APP** $+$ EAC is conservative over **HA** w.r.t. formulae in Γ_0.

Proof. Suppose A to be an arithmetical formula in Γ_0 such that **APP** $+$ EAC $\vdash A$; then **APP** $\vdash \exists x(xr A)$ (by (iii) of the theorem); hence **APP** $\vdash A$ (by (ii)) and therefore, interpreting **APP** in **HA**, **HA** $\vdash A$. $\square$

5.4. We can improve on the preceding corollary by means of more sophisticated methods:

THEOREM (Renardel, generalizing Beeson 1979; also in outline in Beeson 1979A). **APP** $+$ EAC is conservative over **HA**.

We shall not present the proof here.

5.5. *Embedding* **HA**, **N-HA**$^\omega$ in **ML**$_0^i$, **ML**$_0$. This is routine. Let us assume the variables for distinct types be kept disjoint. We identify the types of **N-HA**$^\omega$ with special types of **ML**$_0^i$ or **ML**$_0$

$$0^\wedge \equiv \mathbb{N}, ((\sigma)\tau)^\wedge \equiv \sigma^\wedge \to \tau^\wedge, (\sigma \times \tau)^\wedge \equiv \sigma^\wedge \wedge \tau^\wedge;$$

for terms we put (assuming the notation of Troelstra 1973 for **N-HA**$^\omega$)

$$0^\wedge \equiv 0, S^\wedge \equiv \mathbf{suc}, \Pi^\wedge \equiv \mathbf{k}, \Sigma^\wedge \equiv \mathbf{s}, R^\wedge \equiv \mathbf{r};$$

and for variables $(x^\sigma)^\wedge \equiv x^{\sigma\wedge}$.

For formulae we define a mapping $^\wedge$ to type-expressions by

$$[t =_\sigma t']^\wedge \equiv I(\sigma^\wedge, t^\wedge, t'^\wedge);$$
$$[\forall^\sigma A(x^\sigma)] \equiv \Pi x \in \sigma^\wedge [A(x^\sigma)]^\wedge;$$
$$[\exists x^\sigma A(x^\sigma)]^\wedge \equiv \Sigma x \in \sigma^\wedge [A(x^\sigma)]^\wedge;$$
$$[A \to B]^\wedge \equiv A^\wedge \to B^\wedge; [A \wedge B]^\wedge \equiv A^\wedge \wedge B^\wedge.$$

This gives at the same time an embedding of **HA**, if all variables and equations of **HA** are given the type 0. Regarding **E–HA**$^\omega$ as the theory obtained from **N-HA**$^\omega$ by adding the extensionality axiom at all finite types, we have the obvious

5.5.1. THEOREM. For sentences A, suitable terms t

(i)   if **N-HA**$^\omega$ ⊢ A, then **ML**$_0^i$ ⊢ $t \in A^\wedge$,
(ii)   if **E-HA**$^\omega$ ⊢ A, then **ML**$_0$ ⊢ $t \in A^\wedge$.

5.5.2. COROLLARY. If **HA** ⊢ A, then **ML**$_0^i$ ⊢ $t \in A$ for suitable t.

6. REALIZABILITY AND LOGIC IN **ML**$_0^i$, **ML**$_0$

6.1. *Embedding* **ML**$_0^i$ in **APP**. We now obtain an interpretation of **ML**$_0^i$ in **APP** by thinking of all objects as belonging to a single domain with partial application.

* is a mapping associating with each type-expression $B[x_1, \ldots, x_n]$ of **ML**$_0^i$ a definable comprehension term $B^*(x_1 \ldots x_n)$, definable as a predicate in **APP**, as follows: first we define * for terms

$$(0)^* \equiv 0, (\mathbf{suc})^* \equiv \mathbf{suc}, (\mathbf{k})^* \equiv \mathbf{k}, (\mathbf{s})^* \equiv \mathbf{s}$$
$$(\mathbf{r})^* \text{ is the defined recursor of } \mathbf{APP}, (\mathbf{e})^* \equiv 0,$$
$$(\mathbf{j})^* \equiv \mathbf{j}, (\mathbf{j}_1)^* \equiv \mathbf{j}_1, (\mathbf{j}_2)^* \equiv \mathbf{j}_2; (tt')^* \equiv (t)^*(t')^*.$$

To each variable x of type A we assign another variable in

APP; if we have kept variables for distinct types distinct, we can take $x^* \equiv x$.

For the type-expressions we now put

$$(\mathbb{N})^* \equiv \mathbb{N}, \quad I(A, t, s)^* \equiv \{0 : t^* \in A^* \wedge \ t^* = s^*\},$$
$$(\Pi x \in A.B[x])^* \equiv \{y : \forall x \in A^*(Eyx \ \wedge \ yx \in B^*(x))\},$$
$$(\Sigma x \in A.B[x])^* \equiv \{\langle x, y \rangle : x \in A^* \wedge y \in B^*(x)\}.$$

6.1.1. PROPOSITION. If $\mathbf{ML_0^i} \vdash \Gamma \Rightarrow t \in A$, then $\mathbf{APP} \vdash \Gamma^* \Rightarrow t^* \in A^*$.

Let B be a formula of $\mathbf{N\text{–}HA}^\omega$, $B^\wedge$ its translation as a type in $\mathbf{ML_0^i}$, B^+ its embedding in **APP**; then

6.1.2. PROPOSITION. For suitable terms φ_B, ψ_B

$$\mathbf{APP} \vdash x \in (B^\wedge)^* \to \varphi_B(x)\mathbf{r}B^+; \quad \mathbf{APP} \vdash x\mathbf{r}B^+ \to \psi_B(x) \in (B^\wedge)^*$$

6.1.3. THEOREM. $\mathbf{ML_0^i}$ is conservative over **HA** (under the canonical embedding).

Proof. Let $\mathbf{ML_0^i} \vdash t \in A^\wedge$, then $\mathbf{APP} \vdash t^* \in (A^\wedge)^*$, hence $\mathbf{APP} \vdash \exists x(x\mathbf{r}A^+)$ (combining the preceding propositions), so $\mathbf{APP} + \mathbf{EAC} \vdash A^+$; but $\mathbf{APP} + \mathbf{EAC}$ is conservative over **HA** (5.4) and A^+ is equivalent to A. $\qquad\qquad\square$

6.2. DEFINITION. A type-expression A is $\exists$-free if A does not contain subtypes $\Sigma x \in A.B$ except for cases where B does not contain x free. A term is $\exists$-free iff the types of its subterms are $\exists$-free.

A type-expression C of $\mathbf{ML_0^i}$ is in Γ_0 if for all subtypes $\Pi x \in A.B$ A is $\exists$-free.

6.2.1. LEMMA.

(i)   Let A be a formula of **HA** or $\mathbf{N\text{–}HA}^\omega$; if A is in Γ_0, then so are its embeddings $A^\wedge$, A^+, where A^+ is the standard embedding in **APP** described in 5.2.

(ii)   For $\exists$-free A of $\mathbf{ML_0^i}$ there is a suitable closed $\exists$-free term t_A such that $\mathbf{ML_0^i} \vdash x \in A \Rightarrow t_A \in A$.

(iii)   If B is a formula of $\mathbf{N\text{–}HA}^\omega$ in class Γ_0, then

$$\mathbf{APP} \vdash t \in B^{\wedge*} \to B^+.$$

6.2.2. COROLLARY. (to 6.1.1, 6.2.1 (iii)). $\mathbf{ML}_0^i$ is conservative over **HA** for A in Γ_0, i.e.

$$\mathbf{ML}_0^i \vdash t \in A^\wedge \Rightarrow \mathbf{HA} \vdash A$$

for arithmetical A in Γ_0.

Remark. As it stands, this is just a weakening of 6.1.3, obtained by a much simpler proof. However, the argument shows actually more: if $\mathcal{M}$ is any model of **APP** leaving arithmetical formulae essentially unchanged (i.e., A^+ interpreted in $\mathcal{M}$ is equivalent to A, for arithmetical A) then the combination of 6.2.1 (iii) and 6.1.1 shows that truth in the *model* of $\mathbf{ML}_0^i$ derived from $\mathcal{M}$ (using the mapping * into **APP** followed by interpretation in $\mathcal{M}$) implies truth for arithmetical formulae in Γ_0. This shows that the conservativeness for Γ_0-formulae of **HA** has a strong "invariant" character.

Especially interesting are (1) the model of the partial recursive operations already mentioned, and (2) the model of (total) number-theoretic functions with partial continuous application. The latter model can be formulated and proved to be a model of **APP** in elementary analysis **EL**, which is conservative over **HA**. This can then be used to extend 6.2.2 to $\mathbf{ML}_0^i$ with certain forms of Church's thesis, or certain continuity principles added.

6.3. *Embedding* $\mathbf{ML}_0$ *in* **APP**. This is quite similar to the embedding * for $\mathbf{ML}_0^i$-except that now we associate with each type-expression an *equivalence relation* in **APP**.

    * for terms is defined as before.

For type expressions we put:

$$\mathbb{N}^* \equiv \{\langle x, y\rangle : x \in \mathbb{N} \wedge y \in \mathbb{N} \wedge x = y\}$$
$$(I(A, t, s))^* \equiv \{\langle 0, 0\rangle : \langle t^*, s^*\rangle \in A^*\}$$
$$(\Pi x \in A . B[x])^*$$
$$\equiv \{\langle f, g\rangle : \forall xy(\langle x, y\rangle \in A^* \to Efx \wedge Egy \wedge \langle fx, gy\rangle \in B^*(x)\}$$
$$(\Sigma x \in A . B[x])^*$$
$$\equiv \{\langle x, y\rangle : \langle \mathbf{j}_1 x, \mathbf{j}_1 y\rangle \in A^* \wedge \langle \mathbf{j}_2 x, \mathbf{j}_2 y\rangle \in B^*(\mathbf{j}_1 x)\}.$$

$\langle x, y\rangle \in A^*$, or xA^*y may be read as "x and y equally realize A". That is to say, instead of defining "x realizes A", we now define, by induction on the complexity of the types, a realizability-equivalence at each type.

6.3.1. PROPOSITION. For a context Γ, let Γ^* be obtained by replacing all $x \in A$ by $\langle x, x \rangle \in A^*$; then

$$(\mathbf{ML}_0 \vdash \Gamma \Rightarrow t = s \in A) \Rightarrow (\mathbf{APP} \vdash \Gamma^* \Rightarrow \langle t^*, s^* \rangle \in A^*)$$

($\Rightarrow$ to the right is appropriate if **APP** is given a sequent or natural deduction formulation).

6.3.2. LEMMA (parallel to 6.2.1).
   (i) and (ii) as in 6.2.1, but with **E-HA**$^\omega$ for **N-HA**$^\omega$, $\mathbf{ML}_0$ for $\mathbf{ML}_0^i$.
   (iii) Let B be a formula in the language of **N-HA**$^\omega$ (or **E-HA**$^\omega$), and let $(B^\wedge)^*$ be the corresponding equivalence relation in **APP** as defined above. Then for B in class Γ_0

$$\mathbf{APP} \vdash \langle t, t \rangle \in (B^\wedge)^* \rightarrow B^+.$$

6.3.3. THEOREM. $\mathbf{ML}_0$ is conservative over **HA** for formulae of Γ_0, that is for A arithmetical in Γ_0

$$\mathbf{ML}_0 \vdash t = t \in A^\wedge \Rightarrow \mathbf{HA} \vdash A.$$

Proof. Use 6.3.1, 6.3.2 (iii); compare 6.2.2. □

The remark after 6.2.2 may also be transferred to this case, except for the addition of forms of Church's thesis or continuity principles; these are incompatible with choice and extensionality present in $\mathbf{ML}_0$ (cf. Troelstra 1977). Similar to 6.1.3 we have, with a proof much more complicated than 6.3.3:

6.3.4. THEOREM. (Renardel, generalizing Beeson 1979). $\mathbf{ML}_0$ is conservative over **HA**.

6.3.5. REMARK. In Beeson 1980, a model for the theory of ML 82 is given, essentially that of this subsection specialized to partial recursive operations.
   The idea of extensional realizability of 6.3 is found for example in Pitts 1981, 1.6. Independently P. Aczel, in correspondence, observed that Beeson's model could be seen as a form of extensional recursive realizability.
   The extensional realizability cannot be straightforwardly defined for all of **APP**.

7. CONCLUDING REMARKS; OTHER APPROACHES

7.1. If one assumes in addition to the principles of $\mathbf{ML}_0^i$, that there is a single universe containing all objects, and a uniform, partially defined application operation (corresponding to application in the various fitting types on restriction to the subdomains described by the types), one is led to the theory $\mathbf{APP}$; the interpretation of *logic* in terms of basic domains (types) as given by $\mathbf{ML}_0^i$ corresponds exactly to the interpretation given by abstract realizability for $\mathbf{APP}$.

For $\mathbf{ML}_0$, the interpretation of logic corresponds to *extensional* realizability. Extensional realizability excludes certain possibilities: even if we are convinced that all operations considered are recursive, we cannot consistently within $\mathbf{ML}_0$ adopt the corresponding form of "Church's thesis": $\forall \alpha \exists x \forall y \exists z [Txyz \wedge Uz = \alpha y]$ (α ranging over type $\mathbb{N} \to \mathbb{N}$), for this would require the x to be found as an *extensional function of* α, and this is impossible (cf. e.g. Troelstra 1977). In this connection, we note that many characteristic results of constructive recursive analysis (in the spirit of A. A. Markov) depend on "Church's thesis" in an essential way. The interpretation given by Šanin (1958) for arithmetical statements is essentially equivalent to realizability (cf. Kleene 1960).

7.2. *The theory of constructions of Beeson* (1979A). Beeson (1979A) describes an extension of $\mathbf{APP}$ with added predicates for "p proves A" (so the proofs have to be objects of the single universe), as follows. $\mathbf{APP}$ ($\square$) consists of $\mathbf{APP}$ together with a symbol $\square$, such that for any formula A, $t \square A$ is a new formula; and extra axioms

$(\square \wedge)$ $(t \square A \wedge B) \leftrightarrow ((\mathbf{j}_1 t \square A) \wedge (\mathbf{j}_2 t \square B))$,

$(\square \to)$ $(t \square A \to B) \leftrightarrow \mathbf{j}_1 t \square (\forall x (x \square A \to (\mathbf{j}_2 t)x \square B) \wedge E(\mathbf{j}_1 t)$,

$(\square \forall)$ $(t \square \forall x Ax) \leftrightarrow \mathbf{j}_1 t \square \forall x ((\mathbf{j}_2 t)x \square Ax)$,

$(\square \exists)$ $(t \square \exists x Ax) \leftrightarrow \mathbf{j}_1 t \square A(\mathbf{j}_2 t) \wedge E(\mathbf{j}_2 t)$

and "to assert is to prove"

$(\square)$ $A \leftrightarrow \exists x (x \square A)$, $t \square A \to Et$.

G. Renardel (unpublished) showed that $\mathbf{APP}(\square)$ can be modelled in $\mathbf{APP} + \mathrm{EAC}$ (hence by means of abstract realizability) be defining a simple translation * leaving $\square$-free formulae unchanged such that

$$\mathbf{APP}(\square) \vdash A \Rightarrow \mathbf{APP} + \mathrm{EAC} \vdash A^*$$

(thereby simplifying Beeson's original consistency proof). In fact, for □-free formulae one can go effectively back and forth between xrA and $y \, \square \, B$, provided one assumes $t \, \square \, A \leftrightarrow Et \wedge A$ for A prime.

Sundholm (1982) observes that in the absence of decidability for $x \, \square \, A$ (which contradicts **APP** (□)) the original reason for inserting "second clauses" in the interpretation of $\rightarrow$, $\forall$ (namely to obtain a "logic-free" interpretation) has disappeared.

7.3. Goodman (1970) retains decidability for his interpretation of "p proves A", by introducing a level distinction between proofs; in his theories, however, terms are not always defined.

In Note A of Troelstra 1980 a "trivial" interpretation is described in which second clauses are retained, and also decidability of the proof-predicate (the interpretation is carried out for implication logic, but would apply equally well to all conventional formalisms with modus ponens). One cannot expect, however, to model reflection in the form "provability implies truth" (nor completeness: "truth implies provability"). The trivial interpretation was given only to show that in the absense of reflection and completeness, decidability and "second clauses" can be modelled in a trivial way. Cf. also the remarks in 3.3 concerning a trivial reading of Heyting's explanations.

7.4. Celluci (1981) may be regarded as a formulation of formulae-as-types for intuitionistic predicate logic and arithmetic, without Martin-Löf's richer type structure; he considers both weak and strong versions of ∃E.

7.5. A really detailed and full account of the theory of types and its motivation is still missing. It seems to provide at least partial answers to certain awkward questions we mentioned in 3.3, though the situation is still not as perspicuous as one could wish for; also it remains to be seen whether something like the theory of types is the *only* viable way to make sense of Heyting's explanations.

On the other hand, for model-building, a more naive and pragmatic way of treating basic domains suffices; in particular, this seems to be enough to bring out the realizability aspects as we have done here.

Of particular interest seems to us the possibility to base a constructive set theory on the theory of types (Aczel 1978); at present this seems to us to be the constructively best motivated set theory.

Many questions of a more technical nature concerning the theory of types remain, especially in connection with extensions of ML_0, ML_0^i covering more of Martin-Löf's formalisms; we hope to return to these matters in a future publication.

NOTES

[1] We use "interpretation" here as synonymous with "explanation"; Sundholm 1982 reserves "interpretation" for formal translation procedures. We trust this will not cause confusion here.

[2] We should not think of sets extensionally, but rather as being given by explanations of how their elements are constructed (i.e. Martin-Löf's notion of type).

[3] "Basic domain" is reminiscent of Kreisel's "grasped domain", and akin to Martin-Löf's "type". The terminology of "basic domain" and "basic equality" is meant to be suggestive without suggesting too much. A "basic domain" is somehow more primitive than an arbitrary domain; membership of arbitrary domains is explained in terms of basic domains.

[4] One is inclined to say that one can associate with d a canonical proof d^* establishing that d belongs to D; and for $n \in \mathsf{N}$, n^* would be the (mental) construction of n by successively adding abstract units. Do we have to distinguish between d and d^*? This is the point involved in the distinction between proofs-as-processes and proof-objects (the constructions which are the result of proofs-as-processes); cf. 2.3, 2.4 below.

[5] The principal aim of these clauses was to obtain a logic-free interpretation of the theory of constructions, which might yield technical results, similar to the results obtainable e.g. from the Dialectica interpretation.

[6] Also we shall not always give our rules in exactly the same way as Martin-Löf does; we make some changes to facilitate the comparison with realizability.

[7] This is the model studied in ML 75. In view of the fact that Martin-Löf emphasizes that objects can only be handled via their linguistic interpretation, one is tempted to regard this as the intended model.

[8] This was pointed out to us by P. Aczel.

[9] Here it might have been more elegant to treat λ-abstraction as primitive; in an extensional context this is equivalent.

BIBLIOGRAPHY

Aczel, P. H. G.: 1980, 'The Type Theoretic Interpretation of Constructive Set Theory, in A. Macintyre, L. Pacholski, and J. Paris (eds), *Logic Colloquium* 77, North-Holland Publ. Co., Amsterdam, pp. 55–66.

Beeson, M.: 1979, 'Goodman's Theorem and Beyond', *Pacific Journal of Mathematics* **84**, 1–16.

Beeson, M.: 1979A, 'A Theory of Constructions and Proofs'. Preprint no. 134, Department of Mathematics, Utrecht University.

Beeson, M.: 1980, 'Recursive Models for Constructive Set Theories'. Preprint 129, Department of Mathematics, Utrecht University.

Bruijn, N. G. de: 1970, 'The Mathematical Language AUTOMATH, Its Usage, and Some of Its Extensions', in M. Laudet, D. Lacombe, L. Nolin, and M. Schützenberger (eds), *Symposium on Automatic Demonstration*, Springer Verlag, Berlin, pp. 29–61.

Celluci, C.: 1981, 'A Calculus of Constructions as a Representation of Intuitionistic Logical Proofs', in S. Bernini (ed.), *Atti del congresso nazionale di logica*, Montecatini Terme, 1–5 *ottobre* 1979, Bibliopolis, Napoli, pp. 175–193.

Curry, H. B., and R. Feys: 1957, *Combinatory Logic* I, North-Holland Publ. Co., Amsterdam.

Diller, J.: 1980, 'Modified Realisation and the Formulae-as-Types Notion', in J. P. Seldin, J. R. Hindley (eds.), *To H. B. Curry: Essays on Combinatory Logic, Lambda Calculus and Formalism*. Academic Press, New York, pp. 491–501.

Dummett, M. A. E.: 1977, *Elements of Intuitionism*, Clarendon Press, Oxford.

Feferman, S.: 1975, 'A Language and Axioms for Explicit Mathematics', in J. Crossley (ed.), *Algebra and Logic*, Springer, Berlin, pp. 87–139.

Feferman, S.: 1979, 'Constructive Theories of Functions and Classes', in M. Boffa, D. van Dalen and K. McAloon (eds.), *Logic Colloquium 78*, North-Holland, Amsterdam, pp. 159–224.

Gödel, K.: 1958, 'Über eine bisher noch nicht benützte Erweiterung des finiten Standpunktes, *Dialectica* 12, 280–287.

Goodman, N.: 1970, 'A Theory of Constructions Equivalent to Arithmetic', in A. Kino, J. Myhill and R. E. Vesley (eds.), *Intuitionism and Proof Theory*, North-Holland Publ. Co., Amsterdam, pp. 101–120.

Hilbert D., and P. Bernays: 1934, *Grundlagen der Mathematik* I, Springer-Verlag, Berlin, 1968².

Heyting, A.: 1934, *Mathematische Grundlagenforschung, Intuitionismus, Beweistheorie*, Springer, Berlin. Reprinted 1974.

Heyting, A.: 1956, *Intuitionism, An Introduction*, North-Holland Publ. Co., Amsterdam, 1966², 1971³.

Heyting, A.: 1958, 'Intuitionism in Mathematics', in R. Klibansky (ed.) *Philosophy in the Mid-century. A Survey*, La nuova editrice, Firenza, pp. 101–115.

Heyting, A.: 1974, 'Intuitionistic Views on the Nature of Mathematics', *Bolletino dell' Unione Mathematica Italiana* (4)9, Supplemento, 122–134. Also in *Synthese* 27 1974, 79–81.

Howard, W. A.: 1980, 'The Formulae-As-Types Notion of Construction', in J. P. Seldin, and J. R. Hindley (eds.), *To H. B. Curry: Essays on Combinatory Logic, Lambda Calculus and Formalism*. Academic Press, London-New York, pp. 479–490.

Kleene, S. C.: 1945, 'On the Interpretation of Intuitionistic Number Theory', *J. Symbolic Logic* 10, 109–124.

Kleene, S. C.: 1960, 'Realizability and Šanin's Algorithm for the Constructive Deciphering of Mathematical Sentences', *Logique et Analyse* 3, 154–165.

Kleene, S. C. and R. E. Vesley: 1965, *The Foundations of Intuitionistic Mathematics*, North-Holland Publ. Co. Amsterdam.

Kleene, S. C.: 1973, 'Realizability: A Retrospective Survey', in A. R. D. Mathias and H. Rogers (eds.), *Cambridge Summer School in Mathematical Logic*, Springer, Berlin, pp. 95–112.

Kolmogorov, A. N.: 1932, 'Zur Deutung der intuitionistischen Logik', *Math. Zeitschrift* 35, 58–65.

Kreisel, G.: 1965, 'Mathematical Logic' in T. L. Saaty (ed.), *Lectures on Modern Mathematics* III, John Wiley & Sons, New York, pp. 95–195.

Kreisel, G.: 1970, 'Church's Thesis: A Kind of Reducibility Axiom for Constructive Mathematics', in A. Kino, J. Myhill and R. E. Vesley (eds.), *Intuitionism and Proof Theory*, North-Holland, Amsterdam, pp. 121–150.

Kreisel, G.: 1971, 'A Survey of Proof Theory II', in J. E. Fendstad (ed.), *Proceedings of the Second Scandinavian Logic Symposium*, North-Holland, Amsterdam, pp. 109–170.

Läuchli, H.: 1970, 'An Abstract Notion of Realizability for Which Intuitionistic Predicate Calculus is Complete', in A. Kino, J. Myhill, and R. E. Vesley (eds.), *Intuitionism and Proof theory*, North-Holland Publ. Co., Amsterdam, pp. 227–234.

Martin-Löf, P.: 1975, 'An intuitionistic Theory of Types: Predicative Part', in H. E. Rose and J. C. Shepherdson (eds.), *Logic colloquium* '73, North-Holland Publ. Co., Amsterdam, pp. 73–118.

Martin-Löf, P: 1975A, 'About Models for Intuitionistic Type Theories and the Notion of Definitional Equality', in S. Kanger (ed.) *Proceedings of the Third Scandinavian Logic Symposium*, North-Holland Publ. Co., Amsterdam, pp. 81–109.

Martin-Löf, P.: 1982, 'Constructive Mathematics and Computer Programming'. Reports of the Dept. of Mathematics, University of Stockholm, 1979, no. 11. Appeared in L. S. Cohen, J. Łos, H. Pfeiffer and K. P. Podewski (eds.), *Logic, Methodology and Philosophy of Science VI*, North-Holland Publ. Co., Amsterdam, pp. 153–175.

Pitts, A. M.: 1981, 'The Theory of Triposes', Ph.D. thesis, Cambridge University, U.K.

Šanin, N. A.: 1958, 'On the Constructive Interpretation of Mathematical Judgments', *Trudy Mat. Inst. Steklov* **52**, 226–311 (Russian). English translation: *Am. Math. Soc. Transl.* (2) **23** (1963), 109–189.

Scott, D. A.: 1970, 'Constructive Validity', in M. Laudet, D. Lacombe, L. Nolin, and M. Schützenberger (eds.), *Symposium on Automatic Demonstration*, Springer Verlag, Berlin, pp. 237–275.

Sundholm, G.: 1983, 'Constructions, Proofs and the Meaning of Logical Constants', *J. Philosophical Logic* **12**, 151–172.

Troelstra, A. S.: 1973, *Metamathematical Investigation of Intuitionistic Arithmetic and Analysis*, Springer, Berlin, Ch. I-IV.

Troelstra, A. S.: 1977, 'A Note on Non-extensional Operations in Connection with Continuity and Recursiveness', *Indag. Math*, **39**, 455–462.

Troelstra, A. S.: 1980, 'The Interplay between Logic and Mathematics: Intuitionism', in E. Agazzi (ed.), *Modern Logic – A Survey*, D. Reidel, Dordrecht, pp. 197–221.

van Dalen, D.: 1973, 'Lectures on Intuitionism', in A. R. D. Mathias and H. Rogers (eds.), *Cambridge Summer School in Mathematical Logic*. Springer, Berlin, pp. 1–94.

Institut für Mathematische Logik   Mathematisch Instituut
Einsteinstrasse 62   Roetersstraat 15
44 Münster i.Wf.   1018 WB Amsterdam
Federal Republic of Germany   The Netherlands